CHEATS AND DECEITS

CHEATS AND DECEITS

*How animals and plants
exploit and mislead*

MARTIN STEVENS

OXFORD
UNIVERSITY PRESS

OXFORD
UNIVERSITY PRESS

Great Clarendon Street, Oxford, OX2 6DP,
United Kingdom

Oxford University Press is a department of the University of Oxford.
It furthers the University's objective of excellence in research, scholarship,
and education by publishing worldwide. Oxford is a registered trade mark of
Oxford University Press in the UK and in certain other countries

Published in the United States of America by Oxford University Press
198 Madison Avenue, New York, NY 10016, United States of America

British Library Cataloguing in Publication Data
Data available

Library of Congress Control Number: 2015944361

ISBN 978–0–19–870789–9

Printed in Great Britain by
Clays Ltd, St Ives plc

For Audrey and Samuel

PREFACE

· · · · · · · · · · · ·

Most of us will have encountered one form of deception or another in the natural world. It's hard not to because deception is everywhere. It ranges from the relatively familiar sight of a caterpillar camouflaged on a tree, to the way that some orchids elaborately mimic the smell and appearance of female insects. But what's often less well appreciated is the intricacy, extent, and sometimes the extremity of deceit and manipulation used by animals, plants, and even fungi. This is perhaps partly because deception often occurs in many habitats that we are somewhat less familiar with (such as the deep sea or tropical rainforests), and because it frequently occurs in sensory modalities that we are less well attuned to (such as ultraviolet light or ultrasonic sounds). In short, the more we look for deception and trickery in nature, the more we find it. I have spent most of my career working on various aspects of deception, from camouflage in crabs through to mimicry by cuckoos, and the extent that organisms go to in order to manipulate one another still astounds me. In some regards it shouldn't, because biologists have long appreciated that being successful, in evolutionary terms, is about passing on your genes to the next generation, and if that means tricking and exploiting others to do so, then so be it. It is, however, the sophisticated adaptations and intricacy involved in many forms of deception, and the number of forms of deception that exist in nature, that never ceases to amaze me.

This book is about what we know regarding deception in nature (and what we don't). It is about the types of deception that exist, how they work, and the historical context and significance that deception has had, and continues to have, in understanding evolution and adaptation. It is also very much about the modern scientific work that has sought to investigate and understand deception across a wide range of organisms. Each chapter discusses one or two main types or functions of deception broadly covering three main areas: obtaining food, avoiding being eaten, and reproduction. Without wishing to give too much away, Chapter 1 focuses on a couple of specific examples (especially how some caterpillars trick ants into looking after them) to introduce some key ideas and concepts in the study of deception, and to illustrate just how sophisticated it can be. Chapter 2 discusses how animals mimic other species, or aspects of the environment, in order to steal food from others and to capture their prey. This theme is continued into Chapter 3, which discusses the ways in which many species (especially spiders) use deceptive communication signals and stimuli to actively lure prey towards them through a variety of approaches. The book then turns to how animals use deception in order to avoid becoming a meal themselves, starting with various types of camouflage (Chapter 4), followed by how harmless animals mimic other dangerous species so that predators avoid them (Chapter 5), and then how other species use sudden startle displays and other deceptive tactics to cause predators to flee or ignore them (Chapter 6). Chapters 7 and 8 move on to the ways that animals, plants, and fungi trick and manipulate individuals of the same and different species for reproduction. We start with how birds and insects dupe others into rearing their young, before moving on to how individuals manipulate potential partners and rivals in mating. Chapter 9 then brings things back together to highlight some of the key areas and concepts of deception, and looks forwards to what we still need to understand and discover.

There are several processes that are thought by scientists to drive the evolution of deception in nature. Defining key concepts in a subject like this is important if we are to be clear about where different processes occur, and how the type of deception discussed actually works and evolves. However, while important, there is a risk that formal definitions can become overly dry, and I did not want to interrupt the story of how different types of deception work too much by introducing semantics from the outset. Instead, I have tried to find a balance between discussing the varied types and examples of deception and how they work without being too formal, while still outlining what the key concepts are. Beyond the brief description that follows here,

I have generally introduced key terms, including when they may occur in nature, where they are first relevant in the book, rather than adopting a more formal set of definitions and concepts from the very outset in Chapter 1. Of these, perhaps the most familiar type of deception is *mimicry*. Here, an individual resembles another species in some way, such that it can deceive others into considering it as the 'wrong' object type. For example, some species of harmless snake match the banding colours of venomous species to avoid predators—something called *Batesian mimicry*. Conversely, *aggressive mimicry* occurs when mimicry is used in an antagonistic way. For example, some insects resemble the colour and shape of flowers to avoid being recognized and avoided by their prey, which frequently visit real flowers. *Camouflage* is also a widespread and intuitive idea (though much more complex and intriguing than generally realized). Frequently, camouflage involves matching the appearance of the environment, or some specific object in it (such as a dead leaf), so that a predator (or prey item) fails to detect or recognize it. There is one other key concept worth highlighting here too: *sensory exploitation*. This occurs when an individual produces a communication signal that has evolved to be highly effective in stimulating the sensory system of another animal (of the same or another species). In doing so, the individual making the signal can elicit a greater behavioural response from its target than might otherwise have been the case. For example, some frogs produce mating calls that have evolved specific sound properties that strongly stimulate the hearing sensitivity of females, and in doing so increase their chances of mating. Sensory exploitation seems to be a common way that deception arises and works in nature.

In principle, this book could have been organized by another set of ideas or subjects, and we should not think of the topic(s) of each chapter and the notions discussed as isolated from one another. On the contrary, I have tried throughout to indicate common concepts and theories and attempt to link them together. Nonetheless, the main themes of each chapter seemed to be the most logical way to organize things; being the broad types of deception that exist and what ultimately they achieve. It should also be said that the comparative length of certain chapters and extent of the ideas discussed does not mean that some areas are more or less important than others. Instead, they reflect the relative popularity (past and present) of work testing these areas, and no doubt some lack of knowledge on my part too. The same point goes for the taxonomic spread of the examples discussed, although whether deception is more common in some groups of organism is an interesting question (and one discussed in Chapter 9). Finally, a book like this is not designed to cover all known examples of deception. It's more about our fundamental understanding of the subject area and the scientific experiments designed to test those ideas. The examples chosen are those that I personally felt were most interesting and helpful in highlighting key concepts and the work undertaken to test them. In some cases, I could have chosen different examples.

Writing this book was a challenge, but a deeply enjoyable one. Part of that reason was the substantial help and assistance provided by a great many people. First, I would like to sincerely thank Latha Menon, Jenny Nugee, and Kate Gilks at OUP for helping with all stages of the book, providing a range of comments and feedback on the manuscript, answering a plethora of questions from me along the way, and much more. Their help was invaluable and always highly constructive. Beyond OUP, in the first instance it was Tim Caro who gave me the initial push to bite the bullet and write this book when I mentioned the idea to him some time ago. He also kindly read the entire manuscript and gave considerable feedback on an earlier draft. I also owe a great deal of thanks to the following people, who read various chapters for me and gave a range of feedback, knowledge, and general encouragement: Caitlin Kight, Graeme Ruxton, Lina Arenas, Sara Mynott, Emmanuelle Briolat, Jenny Easley, Sam Smithers, Laura Kelley, Kate Marshall, Anna Hughes, David Nash, Tom Flower, and Tom Sherratt. Many other people also provided useful information. In addition, I would like to thank everyone who kindly supplied (frequently donated) images of their work, making it possible to have such a widely and colourfully illustrated book. Last, but certainly not least, I thank my wife Audrey for her constant enthusiasm, encouragement, and positivity throughout this project. As with everything else, I simply could not have done it without her.

CONTENTS

. . . .

LIST OF FIGURES

· · · · · · · · · · · · ·

1

THE BASIS OF DECEPTION
IN NATURE

. . . .

In the meadows of Europe and Asia lives an endangered and beautiful species of butterfly, the alcon blue (*Maculinea alcon*). Females lay small white eggs on plants, on to which the young caterpillars emerge to feed on the flowers. Once a caterpillar has grown sufficiently, it drops to the ground and is discovered by a species of *Myrmica* ant. At this point, most small insects would be attacked and killed as prey, yet remarkably the worker ants instead dutifully carry the caterpillar back to their nest unharmed (Figure 1). Once inside, the caterpillar takes up residence and is fed by the ant workers as one of their own, using the ants' own resources to grow and develop until the point when it has built up the remaining 98 per cent of the required mass it needs to pupate. The whole process can take up to one or even two complete years, and a month after pupating the new butterfly emerges and leaves the nest ready to repeat the life cycle.

Some studies have estimated that 90 per cent of alcon blue caterpillars that leave their host plants successfully end up inside an ant nest. So what makes them so successful? In short, it all comes down to trickery and deception. Each caterpillar deceives and exploits the ants to its own advantage by resembling their hosts' smell and sounds, so

FIG. 1. The alcon blue (*Maculinea alcon*). Top: an adult butterfly with its beautiful blue coloration. Bottom left: the caterpillars hatch and initially feed on the host plant, before dropping to the ground. Bottom right: a worker ant picks up a caterpillar and transports it back to the nest.

Images David Nash

that the ants take the caterpillar and care for it as one of their own young. The alcon blue is not alone in having such a life cycle, with a variety of other species of blue butterfly sharing similar modes of reproduction (although the taxonomy of this group is complex and controversial).[1]

The alcon blue is one of countless species of plant and animal that deceive others for their own benefit. Many early naturalists, including Charles Darwin and his contemporaries, were all too aware that the natural world is not a harmonious place. While we often see instances of apparent cooperation, selfishness and exploitation rule the day (Figure 2). Organisms face a continuous fight to survive and reproduce, and any advantage in obtaining a mate, locating food, or avoiding predators pays dividends. Unsurprisingly then, many animals and plants (and even some fungi) trick, cheat, and deceive each other to their own advantage, just like the alcon blue butterfly, which deceives ants for resources and a safe place to live. Deception is found widely in nature, from insects in the forest canopy mimicking twigs and dead leaves to hide from predators, to fish that attract prey with biolumin-escent lures in the dark depths of the ocean.

In nature, organisms often engage in some form of communication, in which one individual sends a message or 'signal' to another. For example, a male peacock might try to impress a peahen about his high quality as a mate using his gaudy tail feathers as a signal. Scientists often consider deception in nature as occurring when one party, often of another species, exploits a communication system like this in order to create false, exaggerated, or misleading information. Deception should benefit those practising it, but is often costly to the animals being tricked, from lost time or resources such as food, through to a greatly increased risk of death. This book is about how deception works in nature and how it evolves. It is about how and why some spiders mimic ants, many orchids resemble the smell and appearance

FIG. 2. Examples of deception. Top left: a harmless hoverfly (*Eristalis tenax*), mimicking the appearance of a bee to gain protection against predators. Top right: a freckled nightjar (*Caprimulgus tristigma*) using camouflage to prevent detection from predators. Middle left: many spiders use bright colours to lure prey to their webs, such as this banded-legged golden

orb web spider (*Nephilia senegalensis*). Middle right: camouflage is also used by predators to creep upon their prey, like this preying mantis (unknown species). Bottom left: a nest of a fantail warbler (*Cisticola juncidis*), with two parasitic eggs of the cuckoo finch (*Anomalospiza imberbis*), which uses other birds to rear its young. Bottom right: some male animals mimic females to increase mating success. These are Australian giant cuttlefish (*Sepia apama*), with the bottom individual a female, the top individual a consorting male, and the cuttlefish in the middle a female mimic.

Top and middle images Martin Stevens; bottom-left image
Claire Spottiswoode; bottom-right image Roger Hanlon

of insects, various birds lay their eggs in other species' nests, and much more. We will discuss these questions and many others, including numerous and wonderful examples from nature, as well as the clever science and experiments that have led to our current understanding of deception. Ultimately, this book is about what deception can tell us about how species interact with one another, and the processes of evolution and adaptation.

Let's return to our example of blue butterflies and what they can tell us about deception. Ants are among the most numerous organisms on earth, playing a major role in habitats the world over in their varied relationships with countless other species, from acting as predators through to protectors. Their importance in numerous ecosystems is enhanced by the fact that perhaps as many as 10,000 or more other species of insect live alongside and exploit ants. Many such cheats have evolved adaptations to reduce the chance of being attacked and killed by ants, such as mimicking the chemical cues that ants use to recognize one another. Such species are often called 'social parasites' because of the way each individual parasite exploits not just a single ant but an entire colony (we will cover further instances of this in Chapter 7). The butterfly family Lycaenidae is one of the largest of all butterfly groups, comprising species such as the hairstreaks and blue butterflies. Many of the 5,000 or so species have caterpillars that associate with ants in some way, often mutualistically where both the ant and caterpillar benefit.[2] For example, some caterpillars produce sugary secretions that ants feed on, with the ants in return defending the caterpillars from attack against potential predators. Other caterpillars exploit ants directly for resources, an approach thought to exist in around 200 or more species. The most remarkable cases involve the large blue butter-flies (Maculinea, sometimes now called phengaris)[1] found throughout Europe and Asia, including the alcon blue (M. alcon).

Other Maculinea butterflies start life in much the same way as M. alcon, with the caterpillar initially feeding on the flower buds of a host plant before dropping to the plant base to be discovered and taken back to a nest of Myrmica ants. After this, depending on the butterfly species, one of two things normally happens. Either the caterpillar becomes preda-tory, feeding on the larvae of the ants themselves, or it becomes a type of cuckoo. In the latter case, just like its avian equivalent (much more about these in Chapter 7), the caterpillar integrates itself into the nest and is tended to and fed by the ants directly (Figure 3). In some cases, the worker ants are so busy meeting the caterpillar's needs that they neglect their own larvae. In fact, when food is scarce the 'nurse' ants sometimes even kill their own brood to feed them to the caterpillar. These kinds of lifestyle—'cuckoo' or 'predator'—are found in two separate groups of Maculinea, with genetic analyses estimating that the lineages split around five million years ago.[2]

At this point we might ask in more detail why the ants take the caterpillars back to the nest and tolerate them feeding off their own resources or young. That is, why are the ants so effectively deceived? It had long been suspected that many social parasites that exploited ants relied on chemical mimicry to enter ant nests and stay safe, and that blue butterfly caterpillars mimic the chemical profiles of their ant hosts. But it was not until the end of the 1990s that evidence for chemical mimicry by blue butterflies was clearly demonstrated. A study by

FIG. 3. Top: alcon blue caterpillars being tended to by worker ants inside the nest. Sometimes the same ant colony can have multiple larvae to look after. Bottom: a pupa of the butterfly in the nest.

Images David Nash

Toshiharu Akino and colleagues,[3] at the University of Southampton and the Institute of Terrestrial Ecology in the UK, revealed a number of key aspects of *M. alcon*'s deception. In the first instance, they showed that the caterpillars produce chemicals on their body surface, called cuticular hydrocarbons, and that these closely resemble those that the ants use to recognize other workers and their brood. Next, experiments using small glass dummies showed that, despite being of no value or use to the ants, the dummies would be transported back to their nests by ants when treated with chemical extracts of either the ants or the caterpillars. In addition, the chemicals that the caterpillars made were closer to those produced by the specific species of ant that *M. alcon* targets than to other *Myrmica* species in the area. Further evidence collected by Akino and the team showed that caterpillar mimicry apparently has two components or stages. Caterpillars initially synthesize compounds that mimic the ants' odour profiles in order to gain access to the nest. Then, once inside, they acquire additional compounds that refine their mimicry further. This is likely facilitated either by the caterpillars physically making contact with ants and the nest environment, so that the chemicals rub off on to the caterpillar's body, or by further synthesis of chemicals by the caterpillar. Heightened mimicry is important; caterpillars are more likely to die, presumably as a result of host ant discovery, during the initial few days of integration, and so additional protection is necessary to deceive the ants and survive in the long run. The process might also enable caterpillars to start off mimicking the chemical components of more than one ant species, and subsequently refine their resemblance to just one species, or even one colony, once they have been adopted. Ultimately, mimicry by caterpillars, like many forms of deception, is underpinned by selection for increasing discernment from those parties being deceived—in this case the need for ants to recognize intruders. This likely begins as a general defence against any intruders that fail to match the ant colony's odour profile. As successive generations of caterpillar smell more and more like the ants they target, the workers need to evolve more refined recognition mechanisms. This is especially important in small ant-colonies where loss of eggs through predation or resources through cuckoo-like caterpillars can be especially costly.

The outcome of heightened mimicry by caterpillars and improved defences by ants can trigger a process called co-evolution, whereby changes in the properties of one party (e.g. better mimicry) lead to reciprocal evolutionary changes in the other party (e.g. more refined defences). We will encounter this process several times during the book, especially when we deal with the avian brood parasites, such as cuckoos, in Chapter 7. Sometimes, co-evolution leads to increased specialization by parasites because they must evolve more effective and specific mimicry to overcome host defences. The problem with this, however, is that it can prevent successful targeting of other host species, or even other host colonies that live in different geographical regions if colonies have different odour profiles. In regions where parasites occur, selection can favour ant colonies that evolve new odour profiles. This can enable the ants to 'escape' parasitism because the caterpillars no longer closely resemble the host odours and so are easier to detect. Such evolutionary 'arms races' are common in interactions between species, as the deceptive party continuously tries to better outwit increasingly savvy targets. These concepts are illustrated in blue butterflies because arms races have led to a mosaic of evolutionary interactions in geographic regions, with different populations of butterfly and ant trying to get ahead of one another and each heading off in different directions.[4] As a result, ant colonies of the same species can diverge in their chemical signatures when they live in different places, and the local parasites must do the same thing. As the parasite becomes more specialized with regard to local host colonies they also become less effective mimics of ant colonies elsewhere. As such, specialization in deception can be a double-edged sword because it can close off other opportunities for cheats.

Chemical mimicry is not the only trick that the caterpillars use. It had been reported for some time that they also make sounds resembling the noises and vibrations of ants. This, however, was seldom considered an important part of the adoption and integration process, compared to chemical mimicry. That assumption is reasonable because it's widely known that chemical profiles in ants play a key role in recognition of nest mates and intruders. But the noises caterpillars make do seem to have another role inside the nest, as has recently been shown through work by Francesca Barbero from the University of Turin and colleagues from the UK.[5] The ant *Myrmica schencki* is parasitized by *M. alcon*. Queen ants are afforded special status, protection, and care from worker ants, and part of the way that queens achieve this is through making distinctive sounds and vibrations to the workers around them. The caterpillar also makes sounds and vibrations, and these noises are more similar to those of queen ants than workers. It seems that the sounds allow the caterpillar to elevate its status in the nest, being treated not simply as another brood item or worker, but rather more like a queen. They cause more worker ants to stand on guard and defend the queen or caterpillar, and if such sounds denote special status then this might also explain why some worker ants prioritize feeding butterfly caterpillars over their own brood. This aspect of the caterpillar's trickery demonstrates another broad aspect of deception: that many animals not only have ways to defeat the detection and recognition mechanisms of those they exploit, but once they have done so they often evolve ways to extract as much care or protection as they can. Once pupation to an adult butterfly is complete, the final stage is to leave the nest and emerge. Here, deception is often discarded in favour of a quick escape. The alcon blue, for example, is sometimes recognized as an intruder by ants as it leaves, but is covered in so many loosely attached scales that the ants cannot grab on to it to attack properly.

The utilization of sounds to solicit extra care by caterpillars also allows us to consider for the first time the distinction between two key processes involved in deception that we will encounter throughout this book: mimicry and sensory exploitation. Mimicry is a term often used rather loosely, even by scientists, to describe the situation when two species or individuals resemble one another in some way, for instance through similar coloration or smell. But this basic idea is not strictly accurate, because genuine cases of mimicry occur when one species has come under selection to resemble another closely enough that an observer is fooled into considering it as a different object type entirely. Simply looking the same is not enough. For example, a predatory bird may see a flying insect with black and yellow stripes and mistake it for a nasty wasp, even though it's really a harmless hoverfly. This is called Batesian mimicry, and we will learn much more about it in Chapter 5. In our ant example, one ant may mistakenly recognize a caterpillar as another ant, specifically a queen, and it's this misclassification that characterizes mimicry.

However, there are other reasons why two species or stimuli may come to resemble one another. The most obvious is convergent evolution. For example, both sharks and dolphins have a sleek hydrodynamic shape to move efficiently through water, yet clearly neither animal is trying to mimic the other. Instead, they have a common selection pressure owing to their shared environment. Sensory exploitation is a process often associated with deception, and in some cases it can lead to convergence in the appearance of different species when they share the same observer (for example the same predator species). It is not a simple concept at first consideration, but it's likely to be important in many aspects of deception. In the case of the blue butterfly, the caterpillar produces sounds that solicit extra care. This could be mimicry, but instead the sounds might simply exploit 'preferences' or biases in the way that the ants' sensory systems and behaviour naturally favour certain stimuli. For example, maybe the ants' sensory systems are especially good at detecting sounds of specific amplitudes and frequencies, and the queen ants exploit this bias by producing sounds that match the ants' preferences or sensitivity in order to solicit care and enhance their status. In theory other sounds would work too, but the most effective ones are those that exploit the peak sensitivity in the ants' ability to detect sounds and vibrations. This is just like the way that many begging bird chicks have bright mouth colours and vocalizations that are effective at inducing their parents into bringing more food. Their elaborate displays stimulate the parents into higher provisioning rates. In the ants' nest, a caterpillar would, therefore, benefit by producing sounds similar to those of the queen to extract maximum care from its hosts. Different sounds would be less effective in stimulating the worker ants, leading to reduced care. In this case, the queen and caterpillar may evolve similar sounds not because one benefits from being misidentified as the other, but rather because both have, over time, independently adopted a particular sound that elicits a strong response from the workers. The ants are not making any distinction between the two objects in this whole process (they are not deciding whether the sound denotes 'queen ant' or 'caterpillar'), but merely responding to a stimulus that they favour, leading to greater levels of care. If this is true, then there has been no selection for the caterpillar to mimic the queen ants and to cause recognition errors.

The above may sound far-fetched at first, but numerous studies have now shown that animals have sensory systems with hidden or latent biases towards certain stimuli (such as colours, sounds, or smells) that evolved before any communication signal existed to exploit them. These can arise simply as a by-product of the way that nerve cells making the sensory system are linked together, or they could evolve under selection in another context entirely. For example, some primates (including humans) have very good colour vision for detecting red objects and discriminating these from green backgrounds. This ability most likely evolved under selection to detect ripe red and yellow fruit against green leaves in forest habitats. Since then, red face color-ation has arisen in many of these primate species in mating and dominance interactions, partly it seems to exploit the pre-existing 'preferences' in the sensory system towards the colour red. Because finding food is very important, and the key foods of many primates (ripe fruit and young leaves) are often red or yellow, evolution has shaped the eyes and brains of some primate groups to be very good at detecting red and yellow things. As a consequence, if you could choose a signal to attract a potential mate's or rival's attention, it would make sense to tap into this and make the signal red: that seems to be what evolution has done in these primate social signals. In plants and animals, similar signals have evolved that are effective at exploiting potential mates, prey, pollinators, or predators through strongly stimu-lating their sensory systems and exploiting behavioural biases. We will encounter many such examples and discuss these processes in more detail later on.

In the ants, resemblance could, therefore, arise either by the cater-pillars and queens exploiting the worker ants' sensory and behavioural preferences for certain stimuli (sensory exploitation), or if the ant workers mistakenly classify the caterpillar as an ant queen (mimicry). Perhaps the latter is more likely, because the workers increase guarding behaviour to both the caterpillars and queens in response to their noises, suggesting that they categorize the two things as the same object type that needs protecting. The standing guard response of the workers is apparently not just an elevated response to a favourable stimulus, but rather a particular type of response that implies that the workers consider the caterpillar specifically as a queen. If they responded to the sounds by simply feeding the caterpillar more, then this might suggest sensory exploitation is primarily involved. Predatory caterpillars have also been found to produce sounds resembling queens, and unexpectedly this is just as close in acoustic resemblance by cuckoo butterfly species.[5] This again suggests that acoustic mimicry is more likely than sensory exploitation because the predatory caterpillars do not solicit food from worker ants, but instead binge feed periodically on the ant larvae. At those times they may be especially vulnerable to attack, and so being mistaken for a queen ant may be a good ploy to gain protection. Nonetheless, sensory exploitation may often be a more parsimonious explanation than mimicry for similarity in appearance among species. In fact, some scientists have argued that it could act as a precursor for true mimicry to evolve.[6] For example, resemblance may have initially arisen via sensory exploitation by the caterpillar, helping to ensure that it's well attended to by ants, but then led to mimicry later if the ants came under selection to discriminate between queens and caterpillars.

Returning to the caterpillars, predatory blue butterflies also mimic the odours of the host colony sufficiently not to be attacked. However, they live a stealthy life, not openly flaunting their deception but rather finding a relatively safe place in the nest to lurk before emerging to feed on ant larvae periodically. In contrast, cuckoo species live much more in the open and are capable of extracting significant amounts of food and care from their hosts. A single ant nest can contain six times more cuckoo caterpillars than predatory ones.[7] The cuckoo strategy is also more specialist, and seems to have evolved independently twice during the evolution of large blue butterflies and their close relatives.[2] Being a specialist can enable high success with the right host. For instance, *M. alcon* caterpillars have been estimated to have a chance of success that is thirty times greater when found inside their primary host species' nests than when taken back to the nests of other ant species. The cost of specialization in cuckoo species, as we discussed earlier, is that such open exploitation requires an extremely good level of mimicry to ensure that the worker ants in close contact with them do not figure out that something's amiss. This can constrain cuckoo species to exploit a specific ant species more closely, effectively limiting their options for utilizing other hosts. While predatory species also specialize, this seems to be less specific, and so they can exploit a wider range of ant species. So during deception animals might take a number of different routes, either evolving high specialization on one or a few species or being a jack of all trades generalist, exploiting various species but not being ideally suited to any.

The remarkably advanced and specialist way of life led by blue butterflies is unfortunately also a problem for their survival. Today, many blue butterfly species are endangered, and a loss of habitat and suitable hosts is part of the reason. Their high degree of specialization is likely to have contributed to this downfall. In fact, birds such as some cuckoos, which lay their eggs in the nests of other species, are also more likely to become extinct than birds that show parental care to their own young. Deception and exploitation can be a fruitful way of life, yet the process of specialization can send species down an evolutionary route that leads to a dead end, especially if the exploited species fight back sufficiently strongly and no alternative hosts can be found.

I have used the blue butterflies as an example to begin this book, partly because of their amazing life history but also because they illustrate some common themes that we will encounter throughout our look at deception. This example also highlights another important issue when exploring deceit: human subjective perceptions are often inadequate to study how other animals communicate and trick one another. The sounds and vibrations made by ants and caterpillars are subtle and travel just a few centimetres in the nest. Our ears are not attuned to detect them and so we need specialist equipment to listen in and find out what's going on. Likewise, our sense of smell is far too crude to be able to assess the chemical profiles of ants and the butterflies. It would simply be impossible for us to detect and in any meaningful way analyse the level of mimicry by the caterpillars with our sense of smell, let alone determine differences between colonies of ants. Yet the ants have sensory receptors on their antennae and processing mechanisms in their tiny brains capable of analysing the finest details of the cuticular hydrocarbons on their nestmates and potential rivals and threats. Their sensory system has been shaped through evolution by the ants' ecology and way of life, including the requirement to detect small differences in chemical profiles. The arms race between caterpillar and ant is on a level we cannot perceive. We need to consider how deception works not in terms of our own senses and perceptions, but those of the animals being tricked.

It is not just that our sensory systems are limited either; sometimes our own perceptions can be positively misleading. Crab spiders are a sit-and-wait predator, often positioning themselves on flowers and waiting for an unsuspecting pollinator, such as a bee, to come close enough to strike. To humans, crab spiders are often wonderfully well camouflaged against the different-coloured flowers on which they are found (not least because of their ability to change their colour over a period of days). This certainly seems to be the case with many European crab spiders, which rely on camouflage to remain hidden from their prey. However, it's a different story with some crab spiders found in Australia. Like their European cousins, they often look very well hidden on flowers like daisies to the human eye. Yet this is misleading. For a start, insects do not generally see the world in the same way that we do. Many insect pollinators, for example, have visual systems that can readily detect ultraviolet (UV) light and UV colours in their environment. This is valuable because many floral signals are rich in UV patterns, often acting as 'nectar guides' directing the pollinator into the centre of the flower. Remarkably, unlike European crab spiders, Australian species often positively glow in the UV, standing out like a beacon against the dark flower in UV light (Figure 4). At first, this seems like a bizarre phenomenon. Why on earth should a sit-and-wait

FIG. 4. The Australian crab spider (*Thomisus spectabilis*). The image on the left shows the camouflage of the spider when viewed under wavelengths of light that humans can see. The right-hand image shows how the spider vividly stands out when viewed under UV light.

Images Mariella Herberstein

predator advertise itself to prey? The answer seems to be that the spider actively lures its victims instead of hiding from them.

Astrid Heiling, from Macquarie University in Australia, and colleagues conducted experiments with bees and the crab spider *Thomisus spectabilis*.[8] They first showed that the spiders really do stand out very strongly in UV against the flower, and would be visible to bees. Then, they presented honeybees with a choice between two daisies, one with and one without a spider, and recorded which flower the bees visited most. The bees had a clear preference for flowers with spiders. What's more, when the authors applied a UV-absorbing substance to the spiders to remove their UV signals, honeybees then avoided the flowers with spiders instead. What's also interesting about this system is that it is the introduced honeybees which are most susceptible to the spiders' luring signals. When native Australian bees are presented with the same choice of flowers with and without spiders, despite being initially attracted to the spider signals, they are less likely to actually land on spider-occupied flowers than are honeybees, and instead ultimately prefer vacant flowers.[9] This suggests that the native bees have evolved resistance to the crab spider signals, demonstrating again how deceived animals fight back over evolution against trickery.

Heiling and the team suggested that bees have a pre-existing (innate) bias or preference for highly contrasting UV flower signals because these are common in nature, and that the spiders' coloration exploits this in luring the pollinators more strongly. This is then likely to be a clear case of sensory exploitation. As we noted earlier, sensory exploitation does not require that the bee misclassifies the spider as something else (a flower), merely that the signal it responds to is particularly good at stimulating its sensory system and exploiting some type of general preference behaviour. The spider is not directly mimicking a particular flower, but rather has evolved a colour signal to which bees generally respond to strongly. Thus, our perceptions and the ways that we interpret features in the world around us are products of the sensory systems we have and how they are tuned to specific stimuli. Although humans are highly visual animals we are actually blind to many things, including UV light. Furthermore, much deception in nature occurs in other senses, such as acoustics, vibrations, or chemicals, where our sensitivity is quite inferior to that of other species. In short, our sensory systems pick up only a small amount of the information available to other species, and this matters in understanding how deception evolves because deception frequently exploits the specific ways that animal sensory and cognitive systems work.

Modern science has made considerable progress in understanding deception and how it evolves, partly through the use of sophisticated equipment and a better appreciation of animal senses, coupled with a range of clever experiments. However, we also owe a great deal of debt to some of the early pioneers of evolutionary thinking and natural history. While Darwin is frequently among these, he generally takes a back seat in our story because it was principally his contemporaries who laid the groundwork for understanding deception. Chief among these is Alfred Russel Wallace. Wallace was not only one of the great natural historians and explorers of the Victorian era, spending much time travelling and collecting specimens in South America and South East Asia, but he also arrived independently from Darwin at a very similar, albeit less developed, theory of evolution by natural selection. He also pioneered the field of biogeography, and can even be considered as one of the earliest conservationists.[10] Wallace suggested and commented on a wide variety of deceptive strategies by plants and animals, and how they might have evolved, from camouflage through to mimicry, for both attack and defence. He rather amusingly described species using deception: 'They appear like actors or masqueraders dressed up and painted for amusement, or like swindlers endeavouring to pass themselves off for well-known and respectable members of society.'[11] As we will see at various points in this book, it's striking how many examples of deception were not only suggested by pioneers like Wallace, and indeed how they might work, but that many instances of cheating have only been scientifically tested comparatively recently, well over a century after Wallace made his observations. So, with all this in mind, we can now start our exploration of deception in earnest, beginning with how animals cheat one another to obtain food.

2

THIEVES AND LIARS

One of the main challenges animals face is to obtain enough food and nutrition to make it through each day, and to survive long enough to reproduce. Looking for food takes time, and can in itself be costly due to the energy required for searching, not to mention that used by predators for chasing, subduing, and killing prey. Hunting and foraging also carry risks. Searching for things to eat can mean diverting attention away from looking out for dangers like predators, or even just moving through the environment in a manner that attracts unwanted attention. This all means that animals face the challenge of balancing their energy budgets while ensuring that the time, energy, and risks involved in getting food are worth it. Instead, some animals exploit the resources gathered by others, or use deceptive signals to directly entice victims.

One way that animals can balance the risk of being attacked while still searching for food is by using alarm calls to warn others of impending danger. By doing so, they encourage nearby individuals to be more vigilant to any potential threats, and to take cover and hide should the risk be sufficiently high. Provided others reciprocate this behaviour, everyone wins. Alarm calling occurs even between animals that are from different species. For example, small songbirds like European robins emit calls that alert other individuals from various avian species that a predator, perhaps a sparrowhawk, has been spotted. At first, alarm calling might seem risky to the individual calling because it could advertise itself to the predator. In fact, alarm calls may not carry major costs because they have often evolved properties that make the caller hard to detect and localize. Anyone who's wandered through a woodland when birds are nesting might have noticed the high pitched 'seet' alarm calls but found it hard to pinpoint where they are coming from. Alarm calling is often most sophisticated in species that live in groups, especially mammals and birds, whereby individuals take turns to act as sentinels, looking out for danger and calling when a threat comes near. This triggers the rest of the group to drop what they're doing and dive for cover. In meerkats, a smallish desert carnivore, the sentinel stands on an elevated lookout post to gain a good view of danger all around. The advantages of alarm calling to other individuals nearby are clear: they can hopefully escape before the predator comes too close to be a threat, or at least take some sort of evasive action. In addition, both the sentinel and the foragers can benefit because in groups like this many individuals are often quite closely related. By helping other individuals to avoid danger, the sentinel can also aid the spread of its own genes.

Not all animals are so trustworthy in the alarm calls that they make. Individuals of some species get others to do the hard work by exploiting their food and foraging behaviours. This can be done by exploiting the alarm-calling and escape behaviour of groups of animals by using false alarm calls. Cheats that do this could scare off others who have found food and steal it for themselves. It has been suggested by some that birds may use fake alarm calls to trick others. In 1986 Charles Munn proposed that two species of Amazonian flycatchers, which live in mixed-species flocks with sentinels, make deceptive alarm calls to steal food.[1] However, in this example it was not entirely clear that stealing food did not instead involve acts of direct aggression without trickery. For example, more dominant individuals may have simply bullied others into giving up their food. No study had shown that thieves could mimic the alarm calls of other species for this purpose. In the past few years, all this has changed.

Tom Flower and colleagues have studied the fork-tailed drongo (*Dicrurus adsimilis*), a common bird in sub-Saharan Africa. The species is a 'kleptoparasite', meaning that individuals steal food from other animals, although they also hunt for themselves by aerial hawking for insects or capturing small lizards and insects on the ground. For his doctorate at the University of Cambridge, Flower showed that drongos in the South African Kalahari Desert have a cunning way of stealing food from several species of group-living animal.[2] Individual drongos associate with groups of other birds, such as pied babblers and glossy starlings, as well as meerkats. When a member of a group being watched finds a prey item the drongo quickly emits an alarm call suggesting the presence of a dangerous predator nearby. The victim drops its prey and flees, leaving the drongo to swoop down and claim the food for itself. The drongo's alarm call is not just any old call, but often closely mimics the alarms of the species it is targeting (Figure 5). A drongo stealing from a meerkat emits a meerkat alarm call, whereas a drongo facing a pied babbler makes a babbler alarm. Flower conducted field experiments to show how this works. He set up speakers to play calls of drongos to focal groups of babblers and meerkats, and simulate the trickery of drongos, to test how individuals from those focal groups responded. Both meerkats and babblers were fooled by the drongo alarm calls, dropping their food in the process of fleeing for cover. Funnily enough, Tom told me that drongos even make false alarm calls towards humans. On one occasion (using the alarm call of a white-browed sparrow-weaver) this was enough to cause his two-year-old daughter to drop a worm she was feeding to some other birds, and for the drongo to fly down and snatch it up.

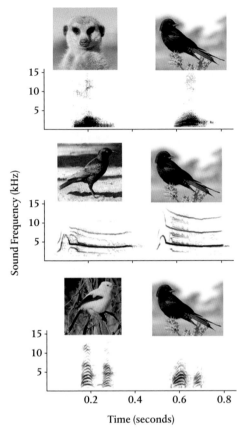

FIG. 5. Multiple deceptive calls of fork-tailed drongo (*Dicrurus adsimilis*). Drongos mimic the alarm calls of a range of target species in order to steal food, including meerkats, glossy starlings, and pied babblers. For each species, the drongo can make a call very similar in frequency and timing to that of the target animal.

Image and figure components Tom Flower

Kleptoparasitism is an important source of food for the drongos, accounting for around a quarter of their diet. It also enables them to broaden their dietary range because they can steal larger terrestrial food items that they might otherwise struggle to capture because other species can dig into the substrate for buried prey. Flower and others have shown that the benefits of stealing food depend on the time of day and weather.[3] For example, kleptoparasitism is more valuable in the mornings and on cold days (mornings in the Kalahari can dip below −10 degrees Celsius), when it's more difficult for drongos to capture prey directly themselves because prey tends to be less active and is often in hiding. In contrast, the species they steal from can still rummage and directly dig them up. The question then is why drongos don't just obtain all their food through stealing; why bother to hunt at all? Well for one thing there are costs and limitations associated with stealing food. Specifically, waiting for others to find food and then trying to steal it is likely to be more unpredictable than self-foraging, where the drongo is more in control. In addition, focal species are aggressive and often fight back against drongos, leading to a risk of injury or at least elevated energy expenditure. So a mixed strategy seems to be the best approach.

There's another reason why mimicry and kleptoparasitism alone is unlikely to be optimal, alluded to by the fact that drongos do not only make false alarms; sometimes they make an honest call when a predator really is present. This is crucial if their deceptive strategy is to work in the long run, because if all drongo calls were false then the exploited individuals should cease to respond and just ignore them. The system would break down if there were not some honest calls to add sufficient doubt in those other species about whether a threat is truly present. So what proportion of drongo calls need to be honest for the false ones to work? This is not a simple question to answer because the relative costs and benefits of the target individual responding incorrectly are different depending on whether the alarm call is honest or not. Furthermore, we also expect individuals to use experience over time to determine how to respond. In cases of mimicry like this the success of deception is often dependent on the relative frequency of the model stimulus being copied. Consider a case where 95 per cent of the drongo calls are false, with no predatory threat. The target individual can ignore these with a high degree of confidence that they are false. Indeed, if they responded each time they would continuously lose their food. In contrast, if a drongo makes alarm calls that are genuine 50 per cent of the time, the targeted individuals are very uncertain about the real nature of the calls because there is a genuine danger needing attention half the time. If a drongo makes sufficient honest alarm calls, the victims need to take note because there could be a predator around. This situation is called a frequency-dependent relationship, and is thought to be very common in many types of mimicry. It arises when the relative cost or benefit of a given strategy depends on its relative frequency compared to potential alternatives. In this instance, the mimicking stimulus (here a false alarm call) cannot be too common compared to the model (the real alarm call) because otherwise it becomes ineffective. In fact, all things being equal, we often expect mimics to be rarer than the model. However, a range of other factors affect the dynamics, including the costs and benefits of making a mistake, which we'll come to in a moment.

The situation is actually a little more complex and sophisticated on the part of the drongo. In addition to mimicking honest and false calls, they also make their own form of alarm call, unique to their species. This use of several different types of alarm call, some genuine and some mimicking different species, may serve to keep individuals that the drongo cheats on their toes. Forcing the focal species (e.g. meerkats or babblers) to remember more calls probably serves to reduce the chance of them learning to ignore the drongos; if the drongo was to make the same deceptive call too often then targeted individuals might quickly ignore it. Consistent with this, when individual drongos target a group they switch between alarm call types. Drongos, therefore, are a fascinating species in which to study mimicry because they use it flexibly. When a hoverfly mimics a wasp, for example, its appearance is fixed for its lifetime and so the relationship between the frequency of occurrence of the mimic and its model (hoverfly and wasp) is at the population level.

In contrast, each individual drongo can be both a mimic (of several different species) and give genuinely honest signals, depending on the circumstance, and so the frequency with which it dupes the target species and how they are fooled is dynamic over even short time periods.

The other reason why we expect focal species to continue to respond to drongo false alarm calls, even when the dishonest ones are common, is that there's an asymmetry in the costs and benefits of responding or not to real and false calls. Responding to a false alarm call and fleeing is costly because it might involve losing a meal, but failing to respond to a genuine alarm call might result in death. So, because the potential cost of ignoring an honest alarm call is much greater than responding to a false one, we would expect many animals to err on the side of caution. What all this tells us is that, in effect, whether an individual should respond to the drongo's deception comes down to a complex equation of the costs and benefits of responding to the different call types, and the relative frequency of honest and false calls over time. Beyond drongos, other bird species, notably Eurasian jays, are thought to use false alarm calls to steal food, but how widespread this behaviour is remains unclear.

Drongos inflict a cost on other animals by stealing food but at least they do not attack them directly, which is more than can be said for some other cheats. This situation arises in another cooperative behaviour, which is again open to exploitation: the mutualistic relationship between cleaner fish and their 'clients'. Cleaner fish work at stations visited by other, often larger, fish that need to have their external parasites and mucus removed. The client gets its parasites taken off, and the cleaner fish gets an easy meal, and so both parties benefit. Cleaner fish are often blue and yellow in coloration to contrast with the reef background and water respectively, making them identifiable and detectable to clients.[4] Bluestriped fangblennies (*Plagiotremus rhinorhynchos*) from Indo-Pacific coral reefs, however, are not so helpful.

They mimic the appearance of juvenile cleaner fish, such as the blue-streak cleaner wrasse (*Labroides dimidiatus*), which as its name suggests has a bright blue stripe running down the side of the body. The fangblennies wait for clients of various species to approach near enough to dart at them, biting a chunk of flesh from the victim in the process (Figure 6). This is a type of deception called aggressive mimicry (a term used by Wallace), whereby a dangerous species, such as a predator, resembles a harmless or beneficial species in order to trick a third party (the client). Aggressive mimics frequently exploit species interactions that are inherently cooperative and in which both

FIG. 6. Disguise and aggressive mimicry by the bluestriped fangblenny (*Plagiotremus rhinor-hynchos*). Top left: a juvenile bluestreak cleaner wrasse (*Labroides dimidiatus*) removing external parasites and mucus from a visiting client fish. Bottom left: the mimetic appearance of the fangblenny when pretending to be a cleaner fish. Top and middle right: fangblennies can also change colour to take on other appearances, sometimes to mimic other groups of fish. Bottom right: fangblennies are equipped with a row of teeth on the top and bottom of the jaw and sharp fangs that disappear into cavities behind the eyes. It is not clear yet whether the fangs are used to attack other fish, or for defence.

Images Karen Cheney

parties benefit, such as clients and cleaner fish. In contrast, animals that gain protection from their mimicry, such as the hoverflies we will discuss in Chapter 5, often exploit antagonistic interactions whereby one party benefits but the other pays a cost. This often occurs between predators and prey, whereby the cost is either death or a lost meal, depending on the outcome. Karen Cheney at the University of Queensland, Isabelle Côté at the University of East Anglia, and other colleagues have investigated the dynamics of the fangblenny and cleaner fish system in detail. This includes observational work undertaken in the field while diving, as well as taking fish back to the lab and conducting controlled experiments in aquaria. In doing so they have uncovered a great deal about how this system works, including some similarities with the dynamics of the drongos and their targets.

Rather than being fixed in appearance, fangblennies can quickly change their body colour in the space of around half an hour, and in doing so adjust the nature of their mimicry.[5] Although colour change might seem an exotic skill, it is actually widespread in nature, from cuttlefish and crabs to chameleons and frogs. It is relatively common in a number of fish groups too. In fangblennies, colour change is somewhat analogous to the different calls that individual drongos can make, and enables them to have alternate strategies. For example, they can adopt colours and patterns very different from the cleaner fish, enabling them to match and blend in with other shoals of fish, which they also attack. Moving individual fangblennies to different situations, in other words around different groups of fish, causes them to change colour, with the appearance they choose being triggered by the presence of a particular fish they like to mimic. That is, fangblennies change to their mimetic colour form when in the presence of a cleaner fish, and change to another appearance when the cleaner fish is removed. Such change is often achieved by special cells in the body called chromatophores (which are found in many animals). These act like little packets of pigment, opening or closing up through either neuronal or hormonal control. When the cells expand, they spread out pigment over the body, such that if the pigment is black, the animal becomes darker. When the chromatophores contract, it has the opposite effect, drawing in the pigment and changing the colour back. In fish, different species use chromatophore cells to either adopt signalling coloration for communication, or camouflaged hues to blend into the environment.

As mentioned already, the fangblenny example differs from that of the drongos because the drongos exploit a system whereby those that they deceive are looking out for a threat (a predator) whereas the fangblennies exploit a relationship where the fish that they deceive look out for a beneficial species (the cleaner). However, like the drongos, the success of the fangblennies is influenced by their relative frequency, in this case compared to both those that they deceive (clients) and those they mimic (real cleaners). Cheney and Côté have found that fangblennies are more successful in attacks when they are comparatively rare compared to the real cleaner fish, and also when there are lots of potential victims around.[6] This fits with frequency-dependent predictions because for aggressive mimics to be successful, they cannot be too common compared to the model, otherwise the victims would quickly learn to avoid them by becoming more cautious in approaching potential cleaner fish and by taking evasive action more quickly. By being rare compared to the potential victims there is also a reduced likelihood that the same victims will encounter frequent attacks, and as a result heighten their vigilance towards fangblennies.

Fangblennies are not only costly to the fish they attack, but also to the real cleaner fish.[7] Cleaner wrasse stations are visited almost 40 per cent less often by clients when fangblennies are present than when they are absent, and clients can learn to avoid cleaner fish and fangblenny associations or even just the area where they work after having a negative experience with a fangblenny. This means that the cleaner fish has a reduction in clientele and obtains less food. The benefit to the client of visiting a cleaner fish also affects the relationship between the three parties.[8] For example, one client, the staghorn damselfish (*Amblyglyphidodon curacao*), is more likely to seek the help of cleaner fish when an individual has a greater infestation of parasites and thus has more to gain from having them removed by cleaners. However, this also leaves them more vulnerable to fangblennies: when the damselfish has more parasites, fangblennies have higher attack success. Clients respond according to the potential risk and benefit of seeking help. When fangblennies are perceived to be rare by clients, and the clients have more parasites that need removing, they should be less cautious and seek help more eagerly. In contrast, if they are harbouring few parasites, and when fangblennies seem common, clients tend to hold off or go elsewhere.

Deception by fish to attack victims is not limited to cleaner fish mimics. Other fish use a different, rather more bizarre approach. One of the most unusual is the tellingly named 'cookiecutter' shark (*Isistius brasiliensis*), which bites pieces of flesh from other fast-swimming species of fish like tuna, swordfish, and even large marine mammals such as porpoises. The shark is a small, slow-swimming tropical species found at a range of water depths, moving towards surface waters at night. It produces green-glowing bioluminescent signals on its underside, which led Edith Widder, who is famous for her work on bioluminescence, to suggest that this mimics the appearance or silhouette of small fish.[9] Many marine organisms when viewed by a predator from the waters below against the light surface appear as a dark, clearly visible shape. A variety of species also emit bioluminescent light from their undersides, often to match the spectrum of downwelling light and hide them with a type of camouflage called counter-illumination. Intriguingly, on the underside of the cookiecutter shark is an area a bit like a collar around the throat and gills that lacks these light-producing photophores found elsewhere on the body, and that is also darkly pigmented. Widder suggested that this dark patch creates the silhouette shape of a prey animal against the general backdrop of downwelling light from above, while the bioluminescence hides the rest of the shark's own body form. The cookiecutter shark may use these deceptive shapes to draw in would-be predators, at which point the shark attacks them instead. The jaw of the shark contains some unusual saw-like tooth shapes, and has lips that act as a suction mechanism to latch on to the surface of an object (like a fish). As a victim passes by, the shark locks on and rotates, pulling a plug of flesh off. Cookiecutter sharks also apparently form groups of multiple individuals, and this might enhance the effect of prey by creating the appearance of a shoal of small fish from below. Although there is no direct evidence that this description of how the cookiecutter shark hunts is accurate, it is consistent with missing crater-shaped chunks of flesh in victims and with similar-shaped chunks of flesh found in the stomachs of the sharks. So why are the sharks not themselves attacked, given that they target larger, fast-moving predators? Widder suggests that the answer might lie in their schooling behaviour, whereby if sharks were attacked the predator would suffer serious wounds by multiple group members, something that Widder describes as being as 'appealing as a swarm of wasps'.

Before moving on from examining how animals trick other species in order to steal food or attack them, there is one more example so remarkable it deserves discussion, in part because it involves several aspects of deception and a complex sequence of events for it to work. Blister beetles (*Meloe franciscanus*) are found widely, including in the south-western US. The larvae of this species hatch and then cooperate to form a dark mass of individuals, sometimes up to 2,000 members strong, congregating on the stem of a plant (Figure 7). This may seem odd, but their 'aim' is to attract a male solitary bee (*Habropoda pallida*),

FIG. 7. Bee mimicry by aggregations of blister beetle (*Meloe franciscanus*) larvae. The mass of larvae creates a ball on the stem of a plant, mimicking a female bee, to which male solitary bees are attracted. When a bee makes contact, the larvae run on to its back. Subsequently, when the male bee mates with a female, the larvae are transferred on to her and taken back to the nest.

Images Leslie Saul-Gershenz

which visits and inspects the larval mass closely, as if responding to the presence of a female bee, allowing the larvae to move on to its back. This occurs very quickly, taking just a few seconds. Later, the male bee finds a female with which to mate, at which time the larvae jump off the male and on to her. It is only then that their sinister intentions become clear. The female bee takes them back to her nest where they then feed on nectar, pollen, and the bee's eggs in the nest.

So how do the larvae achieve this, specifically the important step of getting on to a male bee in the first place? In the first instance, the ball of larvae on the plant seems to mimic the general appearance of another bee, with which the male then tries to mate, allowing the larvae to hop on to his back. To human eyes, this larval mass does not look like an especially good mimic of a bee. However, it is important to consider here that although many bees have good colour vision, like other insects their compound eyes do not provide especially good spatial resolution, especially from afar. Compound eyes in invertebrate animals have hundreds or thousands of lenses or 'facets', each pointing in a slightly different direction, collecting images that are like little pixels that are integrated by the visual system to produce an image. Although a thousand might sound like a lot of samples to gather about a scene, most compound eyes are still not especially good at resolving fine detail about shapes and patterns from a distance. Therefore, a close match for mimicry is not needed when the sensory ability of the animal being duped is limited. Nonetheless, most male bees don't go around trying to mate with other males or trying to copulate with just any old dark round object they find, and so something else must be going on for the bee to be tricked. And indeed it is. Leslie Saul-Gershenz and colleagues from the University of California, Davis have studied blister beetles in the Mojave National Preserve, USA.[10] They noted that male bees often approach the larval balls from downwind, as if following chemical cues. Next they conducted experiments to test this by presenting bees with artificial models of the aggregations, with and without adding extracted chemical cues of the larvae (or just crushed up larvae) (Figure 8). Male bees were more likely to approach the models when the chemical profiles were present than when no chemical cues were added. Furthermore, analysis showed that the chemical cues of the larvae are a close match to the sexual pheromones of female bees.

Blister beetles are a diverse group, with as many as 2,500 species distributed worldwide. Although their biology is not widely studied, many are thought to be kleptoparasites and egg predators of bees, and they often specialize on a small number of host species, using the

FIG. 8. A real aggregation of blister beetle larvae (left), and a model made from aluminium and painted brown. Experiments showed that when treated with the odour of larvae the model also attracted bees, but not when it was unscented.

Images Leslie Saul-Gershenz

adults to transport them back to the host nest. It is not known how this occurs in most species, but the presence of larval aggregations similar to those described earlier, with which male bees try to copulate, and the potential presence of female pheromone mimicry has been described in at least one European species too.[11] So blister beetle larvae, probably of multiple species, lure specific victims by closely resembling their chemical profile (and to a lesser extent visual appearance).

We finish this chapter by discussing an escalation in the cost to the animal being deceived, resulting in its death through deception by a predator. One example bears a number of similarities with the fang-blennies earlier in this chapter. The dusky dottyback (*Pseudochromis fuscus*) is a small predatory coral reef fish found in the Indo-Pacific. It associates with shoals of other fish species, especially damselfish (*Pomacentrus* spp.), attacking and eating the small juveniles. The problem the dottyback faces is that damselfish species can look quite different, including yellow or dark brown in colour (to provide camouflage against different reef backgrounds). Given this, how can the dottyback effectively match individuals from the shoal they are currently with and prevent the damselfish from noticing them? Recent work by Fabio Cortesi at the University of Basel and a range of colleagues, in particular from the University of Queensland (including Karen Cheney), elegantly provided the answer.[12] They built experimental patches of coral reef of different colour, and showed that when individual dottybacks were placed among a shoal of damselfish the dottybacks changed colour over a period of two weeks to match the appearance of their associated damselfish species (brown or yellow) (Figure 9).

However, when the team varied the colour of the background coral, this did not affect the dottyback colour change, showing that the dotty-backs are directing their change towards the target fish, rather than the background appearance. Cortesi and colleagues also analysed the cellular structure of the dottybacks to investigate how they change appearance.

FIG. 9. Dusky dottybacks mimicking the coloration of different damselfish species in order to remain unnoticed while attacking their young. The top image shows yellow damselfish with a dottyback of the same colour below. The bottom image shows a brown damselfish underneath a mimetic dottyback.

Images Karen Cheney

They analysed the appearance of the chromatophore cells, which contain different coloured pigments, finding that yellow dottybacks had higher proportions of yellow pigment-containing cells than black pigment cells compared to the brown dottybacks (Figure 10).

FIG. 10. Chromatophore cells of a dusky dottyback, showing the yellow pigment-containing cells (xanthophores) and black pigment-containing cells (melanophores), which change in relative proportion as the dottyback modifies its colour with time. Scale bar = 100 μm.

Image Fabio Cortesi

The change in colour provides an advantage because dottybacks were more likely to successfully capture juvenile damselfish when the dotty-backs matched the coloration of adult damselfish. It is not known pre-cisely why this occurs, but it is probable that the damselfish simply let their guard down and become less vigilant to threats when there is no obvious sign of a predator. Finally, the dottybacks also gain an added benefit from their deception. Since the damselfish they mimic are camouflaged against the backgrounds where they occur, the dottybacks are correspondingly less likely to be detected by larger predators themselves.

Unlike mimicry, camouflage generally involves blending into the environment in some way, such as a moth resembling the general colour and pattern of a tree trunk to avoid being detected and eaten by a bird. We will focus on this type of prey defence in detail in Chapter 4. However, although less frequently studied, many predators also rely on camouflage to sneak up on their unsuspecting victims, or to lurk and wait for them to come near. This can involve both matching general features of the environment, or more sophisticated mimicry of specific objects. Perhaps the most widely known predators that stalk their prey with stealthy movements and inconspicuous colours and patterns are the solitary big cats such as tigers and leopards. Work in the 1970s pioneered some of the earliest use of image analysis techniques to quantify the patterns of mammalian markings and how closely their size and contrast resembled features of the background. This showed that the structure of tiger markings, in terms of the contrast, size, and frequency of the stripes, effectively blended into the background environment and functioned as camouflage, allowing the tigers to stalk and remain elusive from their prey until it was too late. More recent studies have shown that this is also likely the case with many other species of cat.

The invertebrate equivalents of big cats are jumping spiders, which also often hunt by stealth, equipped with large eyes and excellent vision (despite their tiny size) to stalk their prey. Some species, such as *Portia*, which we'll encounter in Chapter 3, bear a general resemblance to detritus and vegetation to remain hidden from their prey, including other spiders. *Portia* even move with a slow rhythmic rocking motion that helps them to blend into a background of vegetation swaying in the wind. Like many big cats, they also stop and remain still when the prey turns towards them because movement is a major giveaway to animals that are trying to hide.

Stalking prey and relying on camouflage is widespread and effective, but it relies on a good deal of luck because the predator must make all the moves towards the prey in order to capture it without being detected, or just lie in wait for a prey animal to come within striking distance. An alternative is to directly mimic something else in the environment that prey would naturally be drawn to—another type of aggressive mimicry. This is exactly the strategy used by some preying mantis, the alien-looking insects famous for their rapid capture behaviour and remarkable hunting tactics. One species of mantis from the tropical forests of Malaysia, the elusive but widely discussed orchid mantis (*Hymenopus coronatus*), has long been assumed to mimic white and pink flowers, including possessing petal-shaped extensions to their legs and a petal-like body shape (Figure 11). The orchid mantis waits for unsuspecting bees and other pollinators to fly towards it. The species has been known for a long time, and Wallace himself discussed back in 1877 how the orchid mantis may mimic flowers to capture prey:

> One of the last-observed and most curious of these protective resemblances has been communicated to me by Sir Charles Dilke. He was shown in Java a pink-coloured Mantis, which, when at rest, exactly resembled a pink orchis-flower. The Mantis is a carnivorous insect which lies in wait for its prey, and by its resemblance to a flower the insects it feeds on would be actually attracted towards it. This one is said to feed especially on butterflies, so that it is really a living trap and forms its own bait![13]

FIG. 11. The orchid mantis (*Hymenopus coronatus*), which attacks flying prey that venture too close. On the right with a captured insect.

Images James O'Hanlon

However, until very recently this was anecdotal, based on human vision and untested by scientific experiment. This caution is warranted because there are two other non-mutually exclusive explanations for why the orchid mantis looks the way it does. It may simply be camouflaged against the general colour of the flower petals, so that it is not detected at all by prey landing on the plant. Or alternatively it might camouflage itself or mimic flowers to avoid predators, rather than attract prey. If, however, mantis specifically mimic flowers to actively attract insects, then we can make a number of predictions. First, the mantis should match the colour and shape of flowers. Second, they should attract greater numbers of prey than would simply visit their vicinity if they are just camouflaged because prey should be drawn to them. And finally, they should attract prey even when they are not actually on a flower. This latter requirement is important to demonstrate that the mantis does mimic an actual object, the flower, rather than simply being hidden on it. To address these questions and predictions James O'Hanlon from Macquarie University in Australia undertook his PhD on this enigmatic species, and with several colleagues tested its mimicry.[14]

First, O'Hanlon and colleagues got hold of orchid mantis from private Malaysian insect keepers (they are very hard to find in the wild) as well as flowers from locations in the wild where the species is known to occur. Then they used a spectrometer to measure the reflectance of the flowers compared to the mantis. The spectrometer tells us how much light of different wavelengths is reflected from the surface of an object. Broadly, when two objects are close in colour, the reflectance spectra should be very similar. Actually, resemblance is not quite that simple, because what really matters is how the animal in question (in this case a pollinator/prey animal) would view the two objects, based on how those spectra stimulate their visual system, and whether they are capable of telling them apart. So, O'Hanlon and colleagues utilized an already widely used model of animal vision to test this. The model uses the measured reflectance spectra to estimate how the different cells (photoreceptors) in the eye of an animal are stimulated when viewing an object. For example, in humans a reflectance spectrum rich in long-wavelength light tends to stimulate our long-wave (or 'red') sensitive cone cells more than the other cone types, leading to a sensation of the colour red. When the patterns of receptor stimulation between two objects (e.g. a mantis and a flower) are very similar, the model predicts that the visual system cannot tell them apart. The team used a model of honeybee vision, which can see medium-wave ('green'), short-wave ('blue'), and very short-wave UV light. This species' vision was chosen because it is often thought to be representative of the colour vision of many other potential insect prey in terms of the wavelengths of light that can be seen and discriminated (unlike humans, who can see long-wave 'red' light but not UV wavelengths). The model showed that the coloration of the orchid mantis is probably indistinguishable from many of the flowers in the environment to the eyes of a potential prey animal. Whether or not the mantis resembles a specific flower species or broadly mimics multiple species is not clear, but analysis of the coloration and shape of lots of potential flowers suggests the latter is more likely. Indeed, this would seem like a more successful approach because mimicking one flower species alone requires high specialization and might only be successful in places where that flower is common.

Next, O'Hanlon looked at how often insects visited the mantis as opposed to a real species of flower common in the natural environment. To do so, he put wooden poles into the ground and placed on top of them either a real flower, an orchid mantis, or as a control, nothing. He then observed each of these samples for one hour to record how many insects visited them. The results were clear: pollinating insects, including bees, flies, and butterflies, visited the mantis poles more often than they visited bare sticks, and also more than they visited the (live) flowers. On two occasions in the experiment the mantis even successfully caught a prey item. Thus, orchid mantis not only look very closely like a range of flowers, but they attract prey even when not actually sitting on a flower, indicating that their appearance is not just camouflage, but a genuine form of aggressive mimicry. Pollinators detect that the mantis is present (unlike most forms of camouflage, which rely on avoiding detection) yet misidentify the mantis incorrectly as a flower of some sort.

This all seemed straightforward until O'Hanlon did another experiment, in which he made fake orchid mantis of different shapes and colours out of clay and measured prey attraction to them (Figure 12). He found, as expected, that white models (like the real mantis) had higher prey interception rates than brown ones. However, in contrast,

FIG. 12. Models of the orchid mantis made by O'Hanlon and used to test how it deceives its prey.

Images James O'Hanlon

removing the flattened leg structures thought to mimic flower petals or changing the leg orientation to an unnatural position did not affect prey visitation rates. This, O'Hanlon reasoned, suggests that the mantis could attract insects not specifically based on mimicking actual flowers, but rather by exploiting a general preference that pollinators have for objects of certain colours in their environment, a process called sensory exploitation that we have already encountered. The difference between mimicry and sensory exploitation can be subtle, but in genuine cases of mimicry an object is identified by the observer as the wrong type of object. That is, a pollinator sees the mantis but decides it's a flower not an insect. In sensory exploitation, no such (mis) classification process is needed. Exploitation can work simply by an animal being attracted to some stimulus because its sensory (or cognitive) system is especially sensitive to its features. The pollinators' visual systems may just be extra responsive to white objects against green vegetation, and have a behavioural response to go and inspect these stimuli, without going as far as to identify them as anything in particular. We will return to questions of mimicry and sensory exploitation again soon, but for now it is reasonable to say that, although intuitively it looks to us as if orchid mantis mimic real flowers, this might not be quite accurate. Why the mantis have such obvious flower-like extensions to their exoskeleton is unclear if mimicry is not the case, but it could simply be to increase their surface area and therefore chance of being detected, or mimicry could occur at close range when insects are better able to analyse the shape and patterns of mantis and flowers and discriminate between them. That is, initial attraction to the mantis may work by sensory exploitation of prey preferences for generally conspicuous and white objects, but at close range the specific morphology of the mantis may mimic flowers directly.

Animals, therefore, use a variety of deceptive approaches—from sensory exploitation through to mimicry and camouflage—to steal food, attack victims, or even capture prey. The dynamics of these systems and exactly how they work often reflect a complex set of interactions and the relative numbers of each participant involved over time. In some instances, animals have adopted deceptive strategies that can be rapidly changed or modified depending on the current situation, including colour change and using different call types. This can enable them to exploit a wider range of victims and even overcome many of the defences that victims rely on to minimize the costs of being tricked. Most of this chapter has dealt with how cheats gain food or prey through disguise, of either their own identity, or through blending into the environment. However, in many ways this is just the tip of the iceberg, and in Chapter 3 we will deal in much more depth with some of the varied and cunning ways that predators, especially spiders, directly lure their prey.

3

LURED INTO AN EARLY GRAVE

The varied ways that animals steal food from others or hide from their prey that we discussed in Chapter 2 are often sophisticated, but far more sinister methods are also deployed. Some animals, and even plants, actively lure victims towards them, enticing them with colours, smells, or sounds that their prey would often naturally seek out in the habitat. This frequently involves tricking their quarry into the expectation of obtaining a reward, often food, instead of which they are killed and eaten. This chapter is about how animals brazenly lure prey with conspicuous but deceptive communication signals, often involving aggressive mimicry. As we will see, although the range of predators that do this is varied, spiders are the masters of this approach.

We are all familiar with the varied and beautifully intricate webs that many spiders build to intercept flying prey. Indeed, countless species of spider that we regularly encounter sit in the middle of large webs strung between vegetation and other objects. The silk used to build these webs is a remarkable innovation of evolution, being flexible yet extremely strong. It is used by spiders both to physically capture prey and to gather information about types and sizes of prey ensnared in the web through the vibrations the victims make when trying to break free. Webs are even used to transmit information between spiders; many males strum the webs of females in a characteristic pattern in order to entice the females to mate with them rather than eat them.

We could be forgiven for making the assumption that most spider-webs intercept prey through being largely invisible to their victims. Indeed, while doing fieldwork I've blundered headfirst into a big spiderweb more times than I care to remember, so it's easy to imagine that many unknowing victims fly straight into these structures without detecting them too. However, in the same way that predators lie in wait for prey to come close before attacking, building nearly invisible webs and waiting would be relatively inefficient and prone to chance encounters. So, is there anything that spiders can do to increase the prey interception rates of webs? Well, as it happens, there are many tactics.

To begin with, countless web-building spiders entice prey using visual signals. Species from at least twenty-two different spider groups (genera) actively build structures into their webs, often referred to as 'decorations'.[1] These comprise a range of materials, including the remains of dead prey, egg sacs, and other detritus, but also various silk constructions. These silk structures are usually made into lines or zigzag patterns, built either running as a line from the top to the bottom of the web or in the form of a cross ('cruciate'), with four lines running at forty-five-degree angles (to horizontal) diagonally across the web. In theory, web decorations like these could have numerous potential functions, ranging from providing structural support to the web to preventing birds flying through and damaging it by making it more conspicuous. However, evidence indicates that silk decorations are most often used to lure prey. In general, spiders from the genus *Argiope* (comprising around eighty species worldwide) are well known for using silk decorations, with some species using the linear form and others the cruciate diagonal cross type (Figure 13). The St Andrew's cross spider (*Argiope aemula*) is one such species that builds decorations, normally of the cruciate type, and is one of the first species

FIG. 13. The St Andrew's cross spider (*Argiope aemula*) with its bright coloration and characteristic diagonal cross ('cruciate') web silk decorations.

Image Chen-Pan Liao

that scientists used to rigorously test the function of web decorations. In particular, research by Ren-Chung Cheng and I-Min Tso, from Tunghai University in Taiwan, showed that webs with silk decorations attracted more prey than those with the decorations removed.[2]

The silk used in decorations by spiders also differs in its chemical composition to those types of silk used in most other aspects of the webs. One interesting difference is that the silk used reflects larger amounts of UV light. As we have noted, UV wavelengths are invisible to our eyes but can be seen by many other animals, including insects eaten by spiders. The heightened UV component of decorations may enhance their visibility and luring effect towards pollinator species in particular, not least because many flower signals are also rich in UV light. Therefore, silk decorations may exploit sensory preferences in the visual systems of pollinators to lure them to the web, another example of sensory exploitation. Work on another *Argiope* species (*A. keyserlingi*) also found that webs with decorations attract more flying insects than webs without them.[3] In addition, when scientists placed coloured plastic sheets above the webs that filtered out and removed UV and blue light, and in the process prevented the decorations from reflecting UV-blue light, the webs caught fewer bees, wasps, and flies. This suggests that the heightened UV reflectance of silk decorations is indeed attractive to prey. However, this experiment did not also try removing green (medium wavelength) light instead, and so we cannot be completely sure that the reduced insect capture rate was not simply due to the prey avoiding flying into areas with unnatural light conditions.

To investigate whether spiderweb decorations lure prey, we need to consider two key questions: why do web decorations vary in form (linear or cruciate) among species, and how do they work in attracting insects? A particularly comprehensive study in 2010, again by Ren-Chung Cheng and colleagues in Taiwan and Australia, asked exactly these questions of *Argiope* spiders.[4] Animals, including insects, frequently have cells located in their visual system that work to encode levels of symmetry in nature. These may exist because symmetry is common in the world around us in the patterns and body forms of other species, including the appearances of many flowers that insects visit and pollinate. Most flowers are symmetric, and cells in the visual system can encode this symmetry. The cross-like cruciate form presents a particularly strong symmetric signal on either side of a web, which should be more effective at stimulating the visual system of a pollinating insect and luring it to the web than a simpler linear form running down the middle. Cheng and the team used this evidence of symmetry perception in insects to predict that cruciate decorations would be more attractive to flying prey than linear ones. They also argued that linear decorations would have evolved first (presumably as they are simpler in form), followed by the evolution from these of more effective cruciate ones in some *Argiope* spiders. To test this, they used molecular studies of Asian species of *Argiope* to work out how closely related each species is and to construct an evolutionary family tree of the various species. From this, they determined which types of decoration evolved first in the ancestors of the modern spiders, and which type evolved later. As predicted, the linear form does seem to be the original ancestral type, with the cruciate form evolving later in at least two different lineages of the spiders.

Next, Cheng and colleagues performed experiments to test whether the cruciate form does lead to higher prey capture than the linear form. They took webs built by St Andrew's cross spiders and carefully mounted them inside circular wooden frames and presented them to flying prey. As expected, the cruciate decoration did indeed lead to higher rates of prey capture than linear forms. Next they made artificial cruciate decorations from cardboard cut-outs, and rotated these by forty-five degrees so that the decoration lines were running horizontally and vertically on the web, instead of diagonally. These were half as effective at capturing prey than when in the normal form. Therefore, it's not simply the presence of additional lines of silk that matters in luring prey, but rather the actual orientation of those lines.

Overall, it seems highly likely that silk decorations in spiders lure prey through sensory exploitation. This is partly because they seem to exploit pre-existing preferences in insect vision to be especially sensitive and attracted to certain wavelengths of light (such as UV) and to certain patterns of symmetry. While these preferences likely arose in pollinators partly because they regularly correspond to salient stimuli like flowers, it seems unlikely that web decorations are specifically mimicking a flower to work (they do not look much like flowers). So a general sensory exploitation probably explains their effectiveness.

The function of other materials used as web decorations is less clear. However, there is some evidence that detritus, egg sacs, and prey remains are effective in concealing spiders from predators such as wasps.[5] In some cases, the decorations are very similar in appearance to the spiders themselves, so they may act as decoys, drawing the attacks of predators to the decorations instead of the spider itself,[6] although they might sometimes also attract predators too.[7] Therefore, while they may sometimes have a luring or attracting function towards prey, not all decorations do this and they likely exist for other reasons too.

The luring of prey to spiderwebs does not just involve decorations. If you look closely at the coloration of many web-building spiders themselves, especially orb-weaving species found in warmer regions, many are quite brightly coloured and patterned. On first consideration this might not seem a great idea because orb web spiders need prey to fly into their web, and so again we might expect that being hard to see is best for the spider. Yet, in many cases, this is far from what's going on. In fact, just like silk decorations, spider coloration often works as an attractant to flying prey.

Here we again return to *Argiope* spiders. Work in the early 1990s by Catherine Craig and K. Ebert of Yale University investigated the coloration of the tropical orb web species *Argiope argentata* in Panama.[8] This involved placing circular shields or plates covered in grass either in front of or behind spiders in their webs, in order to hide one surface or another, and comparing the amount of prey captured. This revealed some evidence that the rate of prey capture was affected by the spider's coloration, with more insect prey attracted to the high UV-reflecting topsides of the spiders than their undersides. However, the results were far from clear because capture rates of prey overall were sometimes higher when one side of the spider was actually hidden, and the design of the experiment itself—directly hiding the spiders and webs behind physical shields—was far from ideal in that it could have prevented access to the webs by flying prey. More recently, further research by I-Min Tso and colleagues in Taiwan has much more clearly distinguished between two main theories that have been proposed to explain the coloration of orb-weaving spiders: whether the coloration is for camouflage against the background environment, or to lure prey towards the web. The former hypothesis is not as unlikely as it might first seem because many objects in nature, such as flowers, are also brightly coloured, and so the spider's appearance could blend in with those.

In one study, Tso and colleagues investigated the orchid spider *Leucauge magnifica* in Taiwan,[9] a striking species marked with prominent silver and black stripes on its top surface and yellow spots and green stripes on a black background on its undersides (Figure 14). In the first instance, when they removed spiders from their webs, insect capture rates were halved compared to webs left with spiders. This supported the attraction theory, because it's hard to imagine how removing spiders from the web would make the web less camouflaged, whereas it should make the trap less conspicuous. Next, they painted over the spiders' yellow or sliver coloration with green paint to make them less visible, and found that capture rates again declined, supporting the idea that bright body coloration lures prey to the web. Simply putting green paint on the green parts of the spider's body had no effect on insect capture, showing that the process of adding paint itself does not explain the findings (for example, if this changed the smell of the spiders). Finally, the team analysed the coloration of the spiders against the background vegetation, and used mathematical models of insect vision to determine how visible they are likely to be. We encountered these models in Chapter 2 with regard to the orchid mantis. Broadly, they estimate how stimulated the different photoreceptor cells in the visual system of an animal, such as a pollinating fly or bee, would be

FIG. 14. Left: The underside of an orchid spider (*Leucauge magnifica*) with its green and yellow body markings. Right: a giant wood spider (*Nephila pilipes*) with its prominent yellow and black markings.

Images Chen-Pan Liao

when viewing different colours and stimuli. Simplistically, the more two objects (such as a spider and a green leaf) differ in appearance, the greater the differences in the way that they stimulate the visual system should be; if the patterns of stimulation are very similar then an object should be well hidden, whereas if the stimulation is very different then the spider should stand out from the background. In the study here, the silver and yellow colours on the orchid spiders were calculated to be very conspicuous and likely to be readily detectable to a pollinator against the background environment. Thus, the spider coloration likely acts as a visible lure, attracting pollinating insects to the web.

Tso and his colleagues have also studied in detail the coloration and capture success of another species, the giant wood spider *Nephila pilipes*, common in forests in East and South East Asia. Giant is an appropriate descriptor here; it's one of the biggest spiders in the world, with females having a body that can reach up to 5 cm long, and, combined with the legs, an overall size of around 20 cm (Figure 14). Males are tiny in comparison, at just 5–6 mm, and look very different. Unsurprisingly, females sit in huge webs often several metres in size. The spider is marked with elaborate patterns and colours, with yellow and black stripes and spots that look to humans quite conspicuous. Tso and his team have used fake 'dummy' spiders from coloured cardboard cut into the general shape of a wood spider, but with varied appearance (Figure 15). This has revealed that the natural coloration and patterns of the spiders is more effective in attracting prey than either just a plain black dummy or a web with no real spider or dummy at all.[10] The experiment also shows that the coloration of spiders by itself is suffi-cient to attract prey to the webs because, unlike real spiders, the cardboard models cannot release odour cues or other types of attract-ant signals to lure prey. A range of studies by scientists on other spider species has found similar results to those on the giant wood and orchid spiders.[11]

FIG. 15. Artificial dummy spiders made from cardboard to test the role of coloration in attracting prey to the webs of the giant wood spider. The dummy on the left represents the natural coloration of the spiders, which was more successful in attracting prey to webs than the black spider, or webs without spiders/dummies, but less effective than a completely yellow model.

Image Chen-Pan Liao/I-Min Tso

One of the interesting things about prey luring by orb-weaving spiders is that they also capture prey that are active at night. Nocturnal insects like moths frequently have tricks to enable them to see better at night, including modifications to their eyes to capture more light and make their photoreceptors more sensitive. Some insects can navigate and even see colours in the relative gloom. Thus, despite low light, vision is still widely used during night-time hours, meaning that luring behaviour could operate then too. This is exactly what Tso's team have observed in the giant wood spider.[12] They have calculated that, to the visual system of a moth, the spider's coloration is distinctive from background vegetation even under nocturnal conditions. Painting over the markings with black paint (or removing the spider altogether), in much the same way as in their other work, also reduces capture of nocturnal prey. In fact, prey capture rates by wood spiders seem to be even higher at night than during the day. It's a similar story for the orchid spider: painting over their coloration substantially decreases the capture of nocturnal prey such as moths.[13] Indeed, several other species of spider, including the common garden spider *Araneus diadematus* familiar to many people in the UK, have small bright white spots on the undersides of their body that apparently lure prey. Luring of prey to spider visual signals seems to be a very common phenomenon, both during the day and at night.

The key question that few studies have yet to fully address is *why* exactly insects are attracted to the colour patterns of these spiders. One suggestion is that the bright colours resemble specific flower parts or pollen (especially yellow spider colours) in the environment, luring pollinators through mimicry. Indeed, the colour patterns of giant wood spiders have some broad similarities in arrangement to many flowers, although no direct comparison between specific flowers and the spiders' patterns has been undertaken.[14] The mimicry theory might be relatively easy to test experimentally too, and it is perhaps surprising that nobody has yet conducted an experiment that simply adds differ-ent flower petals to webs and measures interception rates of prey. Alternatively, and while not being mutually exclusive from mimicry, insects may simply be attracted to bright colours by virtue of a bias in their visual systems, whereby they are especially sensitive and respon-sive to any bright colours and patterns. In this instance, the spider coloration might not be mimicking specific flowers or vegetation but rather exploiting a general pre-existing preference that exists in the vision and behaviour of many insects. In fact, we already encountered this in Chapter 1, where some crab spiders use glowing UV signals to stand out from the flowers they sit on to lure prey that have a strong preference for UV colours.

Before leaving spider coloration, it's worth noting that having bright coloration to lure insects can come at a cost, in that the spider can become prey itself. In giant wood spiders, the coloration of individuals may reflect a trade-off between attracting prey and risking attack from predators. In the experiments with dummy cardboard spiders, an entirely yellow spider actually attracted more insect prey than spider models with the natural colour patterns. However, these yellow spiders also incurred more attacks from predatory wasps. This nicely demon-strates something that's common in nature: a trade-off between com-peting selection pressures. Trade-offs are widespread in animal coloration and communication signals, and have most effectively been demonstrated as occurring between the demands of attracting potential mates by using gaudy colours while not being too visible to predators. The coloration of male animals frequently comprises this sort of half-way house. The wood spider seems to have a coloration reflecting a trade-off too, being good at luring prey while balancing risks from the spider's own enemies.

In the case of spider coloration and silk decorations it seems most likely that hunting success is enhanced by luring prey via exploiting general biases their quarry have towards certain types of stimuli. These biases can be either inherited ('innate') or learned during an individual's lifetime. However, other spiders use specific mimicry targeted at just a few prey species to get a meal. In many cases, this involves resembling the sexual signals made by potential mates of their prey species to bring prey close enough to attack. Perhaps the most remarkable of these are the bolas spiders (*Mastophora* spp.) from North and South America. In appearance, bolas spiders are not especially interesting, being generally quite small with a chubby abdomen, not unlike many other orb web spiders. But their modest appearance hides a remarkable hunting tactic. They do not hunt by building an impressive orb web like many of their relatives, but instead throw a globule of sticky liquid at passing moths. During the day, many bolas spiders look somewhat like bird droppings to avoid predators (a type of camouflage we will discuss in Chapter 4). However, at night female spiders construct a silk line between vegetation, and then move into the middle of this 'tightrope' where they hang. Next, they assemble another single strand of silk, on the end of which is added a sticky adhesive ball (the bolas). The spider holds this with one of her forelegs and swings or launches it at a passing moth (Figure 16). When she hits a target, the sticky substances penetrate through the insect's scales and snare the moth, allowing the spider to pull in the doomed prey. Anyone who watches a video of a bolas spider hunting will surely be impressed by its effectiveness. But when we consider what the spider is doing, it seems like a rather inefficient way to hunt: simply waiting for a moth to come close enough to attack. Yet early work suggested that the amount of prey that female bolas spiders capture is not dissimilar to other orb web spiders of about the same size that build webs. There's a trick to their success, and it's a very sophisticated one too.

In the late 1970s, William Eberhard from the Smithsonian Tropical Research Institute noticed when studying bolas spiders in Colombia

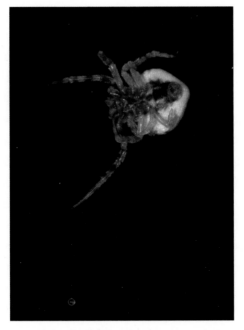

FIG. 16. A female bolas spider (*Mastophora hutchinsoni*) hanging from her 'tightrope wire' and dangling a sticky bolas at the end of a silk line, waiting for a moth to approach.

Image Kenneth Yeargan

that moths did not haphazardly fly around the general vicinity of the spiders, but instead seemed to fly towards them.[15] The flight path comprised a characteristic zigzag pattern, with moths flying from side to side as they approached, in much the same way that male moths fly when tracking a female pheromone odour plume being blown into the air. Removing the sticky bolas did not change the moths' behaviour, so it seemed that the moths were attracted to the spiders themselves. The implication was that bolas spiders are somehow attracting prey by mimicking the sexual pheromones of female moths themselves. About a decade later, scientists analysed the chemicals that were emitted by female bolas spiders and showed that they do indeed mimic specific pheromone compounds of the moth species they capture.[16] Later work also showed that spiders mimic not only the key compounds found in the pheromones of their main prey, but also the ratio in which components would normally be released.

Adult female bolas spiders seem generally to rely on capturing just a few species of moth (and only the males). Kenneth Yeargan and Kenneth Haynes from the University of Kentucky and colleagues identified that for one species of spider (*Mastophora hutchinsoni*), 93 per cent of its diet comprises just two species of moth, the bristly cutworm (*Lacinipolia renigera*) and the smoky tetanolita (*Tetanolita mynesalis*).[17] Individual spiders capture both moth species even though the moths are active at different times of night (the bristly cutworm soon after dark until around 10.30 p.m., and the smoky tetanolita after 11.00 p.m.). This raised the question of how the same spider can lure two different species of moth, which would likely have quite different female pheromones. In theory, the spiders could emit a blend of smells containing components of both moth species' pheromones. However, this may not be accurate enough to lure either species effectively because pheromones tend to be highly specific to each species. Alternatively, spiders could change the components and proportions of chemicals that they emit depending on the time of night. In fact, it seems that both options are partially true—spiders are capable of attracting both moth species outside their normal periods of activity, indicating that there must be some degree of generality in the emissions produced. However, spiders start the evening making relatively more of the pheromone components of the bristly cutworm, and then as the night goes on reduce this in favour of making components that match the pheromones of the smoky tetanolita.

These remarkable adaptations demonstrate how bolas spiders lure prey to them with a strategy called aggressive mimicry, just like that of the fangblennies that mimic cleaner fish (Chapter 2). However, the bolas spiders still need to physically capture the moths, so how do they know when to swing their bolas? Eberhard originally noted that spiders could be induced to draw back the bolas by making a low-pitched humming noise, and later work confirmed that spiders respond to the vibrations of moth wing beats—either to those of real moths or to simulated encounters using sounds played from speakers. So the spiders first lure moths by mimicking the sexual signals of females, then listen to the sounds made by approaching moths to determine when to strike.

The bolas spider is a wonderful example of just how specialist aggressive mimicry can be in some species, and in fact just how specialist it often must be for it actually to work. It also shows how some individuals can change the way their mimicry works over time to improve its success, not dissimilar to the food-stealing drongos of the Kalahari (Chapter 2). Bolas spiders are in fact far from an isolated example in luring prey by mimicking the sexual signals of potential mates. Fireflies are well known for their alluring and beautiful flashing lights, with colours ranging from yellow to a greenish blue. Normally, both males and females produce flashes of light in order to communicate during courtship and attract one another for mating, with the colour and the pattern of light flashes often varying from one species to the next. Males usually start the process with a relatively elaborate pattern of flashing, and then wait to detect the particular response flashes by females of the correct species, often denoted by the length of the delay before the female responds with her own flash. However, some fireflies from the genus *Photuris* use their flashes in a deceptive manner. Females are 'femmes fatales' fireflies, being predators of other normally unrelated species (such as *Photinus*), and they mimic the bioluminescent signals (time delays) of females of their target species to attract males, which they then eat.[18] The twist in the story is that to the predatory *Photuris* females, it's not just about getting a meal. In fact, they are also after particular chemicals, called lucibufagins, that their

Photinus prey species produce.[19] These defensive chemicals repel predators such as spiders and birds, which learn that the flashes of some fireflies are a warning that they are toxic to eat and should be avoided. The predatory *Photuris* fireflies don't seem able to make the chemicals themselves, but they can obtain them from their prey, thereby getting a meal and deriving protection from predators from the same behaviour.

Perhaps the most cunning of all spiders is not one that sits and waits for its prey, but an active hunter. The jumping spider *Portia fimbriata*, a species found in Australia, eats other spiders. It has a repertoire of tricks that have been studied by Robert Jackson from the University of Christchurch in New Zealand and colleagues. It first relies on cryptic appearance and behaviour to sneak up on its prey. In form, *Portia* resembles detritus to blend into the background, but it also has a characteristic mode of walking that Jackson and others have termed 'cryptic stalking', whereby the spider walks towards its prey in a slow mechanical way, hides its palps from view, and freezes whenever its prey turns to face it. Like other jumping spiders, it has large eyes and acute vision. One of *Portia*'s main prey, another jumping spider called *Euryattus* found in Queensland, makes a nest inside rolled-up leaves suspended from rocks or vegetation, and *Portia* must lure it out in order to attack. Male *Euryattus* court females by standing on the leaf nest and 'shuddering', causing the nest to move and rock, drawing the female out. *Portia* descends on to the leaf, and once on top makes vibratory signals resembling these courtship displays, enticing the females.[20] While waiting for the female to emerge, *Portia* lurks within a few millimetres of the nest entrance, sometimes for many minutes, before shuddering again if the female does not appear. *Portia* do not seem to ever enter the nest of *Euryattus*, perhaps because their prey is also a formidable predator in its own right, and this would simply be too risky.

Two further things are interesting about *Portia*'s behaviour. First, *Portia* spiders from localities where *Euryattus* is not found don't utilize this courtship-mimicking behaviour. This suggests that the tactic used by the Queensland population of *Portia* is a trick used to entice a specific local prey species. The behaviour occurs in both wild-caught *Portia* from Queensland, and individuals descended from the same population but reared in the lab, and so without prior experience of *Euryattus*. This tells us that the behaviour is probably genetically encoded or 'innate' to this population, rather than learned through hunting experience. Second, *Euryattus* spiders seem to recognize the approaches of *Portia* as those of a potential predator. Thus, *Portia* most likely needed to evolve their deceptive signals to overcome the defensive behaviour of *Euryattus*. Even so, when *Euryattus* is under attack individuals often charge or leap at *Portia*, or hit them with their legs to drive them away. This defence is often successful, at least in cases where the two spiders have been allowed to interact in lab experiments. In fact, the defence seems most likely to work when *Portia* is approaching the nest, perhaps because once *Portia* is on the leaf and making the false mating calls it has often reached its position undetected.

As if these hunting tactics were not impressive enough, *Portia* has other tricks up its sleeve. It also targets other spiders that live in webs (which have poor eyesight and rely more on vibratory signals to recognize prey and danger).[21] Here, *Portia* strums the silk threads of the web to manipulate the behaviour of the resident spider, either to lure it to within striking range or to mislead it in some other way. Remarkably, the exact tactic used depends on the size of the prey species. When the resident spider is small, and therefore not a major risk, *Portia* creates relatively strong vibrations resembling those of a captured insect, luring the resident spider towards it. However, when their prey is large, and thus could attack and kill *Portia* itself, *Portia* creates more subtle ambiguous vibrations that broadly simulate the presence of some unknown creature at the outer reaches of the web, keeping the resident spider in the middle of the web. This also reduces the ferocity of the targeted spider's attack behaviour, instead promoting reduced inspection behaviour (often involving the prey species plucking and probing the web and slowly moving towards the source of the ambiguous signals). In this way, *Portia* can stalk it with less risk to itself. Finally, when *Portia* is moving towards a spider in its web, irrespective of the spider size, it uses a third type of deception. It makes brief but strong movements that simulate a large-scale action on the web, such as a falling object like a leaf landing on it or the web being blown in a gust of wind. This 'smokescreen' hides the vibrations created by *Portia* as it moves, so that they are ignored by the resident spider, enabling *Portia* to creep towards its target. All in all, *Portia* is one remarkably adept predator.

Quite why spiders in particular are so expert at luring prey with deception is not entirely clear, nor why such strategies appear less common in other animal groups. However, what seems apparent is that aggressive mimicry for prey acquisition often evolves in predators that broadly sit and wait for prey to come close enough to attack. As we discussed already, such a hunting approach often relies on a certain amount of luck. Active luring in these species, especially ones that specialize in capturing a limited number of prey types, is a powerful way of increasing success. It's not just spiders that engage in this approach either. Some predatory prawns lure fish with colours that exploit their prey's preference for certain colours. For example, one species of prawn from Trinidad attacks guppies (the type of fish often found in pet shops with males having gaudy coloration) and has orange spots on its pincers. Guppies tend to have preferences (sensory biases) for the colour orange, especially as this is an important signal during mating and perhaps foraging, and experiments have revealed that the orange spots on the prawns attract female fish to the front of the body where they can be attacked more easily.[22] Cases of luring by vertebrates are generally rarer but also seem to fit this general observation. For example, a number of snakes apparently use luring to attract prey. The death adder (Acanthophis antarcticus), for example, wriggles or twitches the end portion of its tail, while otherwise remaining still, to lure victims through tricking them into responding as if the tail could be a prey item itself.[23] But the vertebrate masters of luring and deception for hunting are a group of fish.

The deep ocean might seem like vast, dark, empty space, but that's only half of the story. If we were to descend down into the water depths, beyond the last rays of sunshine, flashing lights begin to appear, belonging to a plethora of weird and wonderful creatures. Many of these, such as octopus and jellyfish, often look broadly similar to their shallow-water cousins. Others are strange and look like nothing much else. On the ocean floor we might come across a small moving blue-green light, swaying back and forth like a glowing creature moving in the depths, perhaps foraging on dead material that has rained down from above. A fish may be drawn closer too, perhaps considering the glowing object as potential prey. But when it gets close there is an explosion of sediment and a blur of movement, and the fish is gobbled up by a huge set of teeth-filled jaws. The real predator is an anglerfish, one of the most infamous animals known to attract prey with trickery. Currently, there are over 320 species of anglerfish recognized, with about half of these found in the deep sea below 300 metres or so.[24] The general group they belong to includes species such as monkfish, well known for their delicious meat, and the strange looking frogfishes, with different species being diverse in both body shape and coloration. Some can appear bizarre, such as the hairy frogfish (Antennarius striatus), which is round in shape with a huge mouth, and covered in hair-like extensions to camouflage it from its prey.

Despite the challenges involved in studying them, we know that many anglerfish have remarkable adaptations to cope with their environment. For example, deep-sea species face a major problem in finding mates because the limited food and vast spatial scale means that relative population densities are often very low. This makes it a rare event for a male and female to come together at the right time to mate. Some species have overcome this with an ingenious, and frankly bizarre, solution.[25] The male is very small, a strategy known as dwarfism, whereas the female in comparison remains huge, often several times his length. The male searches for a female, often using a highly developed sense of smell from a distance, and then visually when he's closer. Alternatively, in some species the male may zero in on the female's glowing lure (more on that shortly). When the two meet, instead of mating and going their separate ways, the male attaches himself to the female's body, holding on tight with pincer-like apparatus on the tips of his jaws. Over time, he gradually fuses to her. This remarkable process, which is not fully understood, involves both breaking down the skin and tissue barriers where the two individuals touch, and the male linking up with the female's blood supply for nourishment and oxygen, degrading his own digestive systems, other internal organs, and eyes at the same time. Simultaneously, the reproductive organs of the male mature and he essentially becomes a small parasite attached to the female, but one that carries sperm to be released whenever the female intends to spawn. Sometimes, several males can be attached to the same female. Not all species do this permanently—some attach just temporarily without such extreme changes in morphology—but in this way they can overcome the big problem of having to look for a mate.

Anglerfish come in a variety of forms and live in many environments, but it's the modified dorsal-fin spine used to make a lure in many species that gives the group its name (Figure 17). This is a bit like

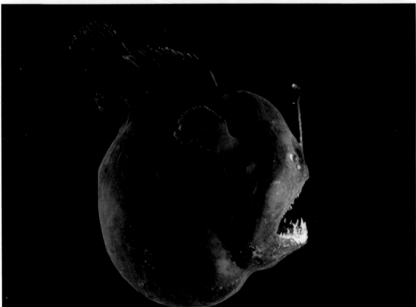

FIG. 17. Two species of anglerfish (*Linophryne arborifera* and *Melanocoetus murrayi*), demonstrating the group's highly diverse morphology, including the characteristic lure used to attract prey.

a fishing rod dangling over the head or body that the creature waves around and flicks about to attract prey, such as other fish, while lying still and remaining mostly hidden. When a prey animal comes close enough the anglerfish engulfs it in a big mouth full of teeth. Many deep-sea species have glowing lures, shining with bioluminescent light created by millions of bacteria in the end of the lure. The bacteria are symbiotic with the anglerfish; they seem to enter from the water into the lure through pores, and then gain nutrients and protection while the fish benefits from the glowing light. Interestingly, there are many species-specific differences in lure shapes and light emissions, and this might also help males to find and mate with the correct species not just to lure prey. The range of remarkable adaptations, including bioluminescent lures and their curious reproductive strategies, has led some scientists to suggest that the diversity of mid- to deep-sea anglerfish stems from these evolutionary innovations. This has allowed anglerfish to diversify quickly and occupy environments that are limited in food and pose challenges for finding mates.

Fish themselves are also victims in the deep ocean, with a rather strange group of deep-sea organisms also seemingly using bioluminescent lures to attract victims. Siphonophores are colonial hydrozoans, a type of organism related to jellyfish, that can reach several metres in length and have tentacles with stinging cells to capture prey. Studying deep-sea species remains challenging, not least because accessing their environment is so difficult, and because bringing specimens to the surface relatively intact is complicated and often not even feasible. Nonetheless, recent work has managed to capture several specimens of siphonophore using submersibles as far down as 1,600–2,300 metres. This has revealed that some individuals have light-producing cells at the end of transparent side branches to their tentacles.[26] These are moved in a rhythmic flicking behaviour, producing yellow and red light, and are likely used to lure fish prey. Deep-sea squid have also been observed via footage from remotely operated deep-sea vehicles using modified tentacles to lure prey, perhaps mimicking small marine organisms.[27] They lack light-producing organs but might attract prey by triggering bioluminescent discharges in nearby organisms, drawing other species to them, or by creating vibration disturbances in the water that are picked up by the sensory mechanical receptors of fish, crustaceans such as shrimp, and other cephalopods. Luring and deception are undoubtedly common in the deep ocean, and if only we had more resources to discover it we could learn a lot about how animals communicate and trick one another.

Deception by animals to lure prey spans the deep ocean to tropical forests. Yet, as if the varied examples of animals engaged in deception were not enough, plants have also got in on the act. While most species of plant are relatively harmless to animals (although see Chapter 8), around 600 species, particularly ones living on nutrient-poor soils, enrich their diet by digesting arthropods (and sometimes larger animals) that they capture. This is most famously known in the Venus fly traps, pitcher plants, and sundews, but carnivory by plants seems to have independently evolved multiple times. Darwin was fascinated by carnivorous plants and spent much time studying them, so much so that he even wrote a book about them, their curious lifestyles, and how they capture their victims.[28] Just like their animal counterparts that are sit-and-wait predators, carnivorous plants have also evolved tactics to lure prey to them and increase capture rates, because simply blending into the background vegetation would not be a particularly effective way of hunting. Like those of some spiders, the methods carnivorous plants use can include both visual and chemical deception. Anyone who has looked at carnivorous plants in detail will likely have noticed that they are diverse and often beautifully marked. Venus fly traps, for example, possess red colours in the centre of the traps, pitcher plants display a wide range of stripes and veins that seem to positively glow at times, and sundews are bright red where the sticky 'glue' is produced. But have these signals really evolved to lure insects and other animals? Recent work suggests that this is indeed the case.

Martin Schaefer from Freiburg University in Germany and Graeme Ruxton from St Andrews University in Scotland compared the capture rates of insects by pitcher plants that were artificially coloured with paints to be either uniform red or green.[29] They found that the red plants caught more insects and suggested that the elaborate patterns and markings found on pitchers might not always be needed for attracting prey. Instead, high contrast against the general green background environment might be sufficient, even of just uniform red coloration, and the various markings carnivorous plants have might simply represent good ways of doing this. However, other work has shown that some insect groups, such as ants, which often make up a significant proportion of certain pitcher plant diets, are attracted to nectar that the plants release, rather than their colour.[30] So there may not be a universal rule that pitchers lure prey with red signals. Indeed, many pitchers are green with white veins.

As it happens, it is not just red coloration that is thought to attract prey anyway. Many pitcher plants have patterns that are highly visible in UV light, which many insects can see. Interestingly, other markings actually absorb UV light and then re-emit this at longer 'blue' wavelengths, a process called fluorescence (Figure 18). It is the same process whereby fluorescent materials can glow brightly when put under a UV-emitting black light, as in some 1980s nightclubs. This blue fluorescence seems to be important because a range of pitcher plant and flytrap species produce beautiful blue fluorescent displays, in particular located around the entrance to the pitchers. When scientists have eliminated this blue glowing signal, prey capture rates to the plants decline substantially.[31] One of the authors of that work, Sabulal Baby, tells me that they have found UV-induced blue

FIG. 18. Pitchers of the plant *Nepenthes khasiana* under white (human visible) light (top), and UV light of 366 nm (bottom).

Images Rajani Kurup, Anil J. Johnson, and Sabulal Baby

fluorescence in all of the 13–14 species of *Nepenthes* pitcher plants that they tested, and in several hybrids. So it is likely that this phenomenon is widely utilized to enhance prey capture.

Carnivorous plants also lure prey with chemical signals.[32] For example, the Venus flytrap releases a cocktail of volatile substances that attracts flies.[33] These plants are famous for snapping shut a trap when an insect lands on it, based on the triggering of special sensory hairs. Chemical analysis of the volatile emissions shows that the plants entice prey by seemingly mimicking some of the odour cues characteristic of foods that flies eat. Pitcher plants also have a similar approach, releasing an abundance of scents with compounds that apparently resemble those found in flowers or fruits to draw in insects.[34] Greater numbers of flies are attracted to these pitchers when the amount of chemical released is higher. Likewise, ants are also attracted to the scents of pitcher plants. What is unclear at the moment is whether carnivorous plants mimic a specific flower or fruit species, or whether they just generally resemble common substances found in those objects. Intriguingly, chemical analysis of some pitcher plant compounds has found substances similar to those used in ant communication, including pheromones used to establish trails and recruit others to food, raising the possibility that the plant chemically mimics an animal communication signal. Finally, and paradoxically, some pitcher plants seem to use a strategy of temporarily deactivating their traps and allowing prey to escape to increase overall levels of prey capture.[35] This might seem odd, but the explanation lies in the fact that ants form a large part of their diet, attracted to the pitchers by sugary solutions secreted from the pitcher rim. When some scout ants which have been sent out to find new food sources escape, they can recruit other individuals from the nest to return later on, thus increasing the pitcher plant's success when the traps are reactivated. So, just like animals, carnivorous plants employ a range of deceptive mechanisms to enhance their diet.

So what do all these examples tell us about deception? For one thing, they show that mimicry can at times be extremely accurate and directed towards just one or two target species (or even a single sex, usually males), as is the case with *Portia*, bolas spiders, and fireflies. But deception can also be more general. Anglerfish do not seem to mimic any specific prey species or type, but rather use a general stylized lure to draw prey in. This again raises an important point about whether all the examples of luring and deception that we observe are truly mimicry (involving misclassification of one object as another), as opposed to simply exploiting the pre-existing preferences that many animals have for certain types of stimuli (that is, sensory exploitation). It is worth dealing with these issues again here because they often reflect different pathways down which deception can evolve.

As we know already, while the difference between mimicry and sensory exploitation can be subtle, they are not exactly the same thing. For example, if a prey animal is very sensitive to the colour blue—perhaps because it's a pollinator and many flowers are blue—a predator wishing to attract its attention may do well to evolve a blue lure. In the case of the anglerfish, the shaking of a lure, especially a bioluminescent one, may just be a very good way of attracting attention and drawing prey animals closer in to inspect it. In the case of the bolas spider, there can be little doubt, however, that mimicry genuinely exists because of the sophisticated use of compounds directly comparable in effect to the female pheromones of prey species. All the same, demonstrating evidence for mimicry as opposed to sensory exploitation is not simple.

Mimicry involves two stages. First the deceptive signal of an animal must be sufficiently similar to the thing it is copying that the sensory system of the observer cannot easily tell them apart as different types of object. As we'll discuss in Chapter 5, for various reasons this does not always require a very close match in all traits. For example, an observer may only pay attention to some aspects of appearance but not others. Second, and following on from this, the receiver must misinterpret the information and identify the mimic incorrectly as the wrong object type. That is, mimicry involves misclassification of one object for something else. A moth is lured to a bolas spider because he erroneously considers the chemicals released as belonging to a female moth.

In contrast, sensory exploitation does not have this requirement for precision in copying and misclassification—it simply involves a signal that's very good at eliciting a response in an observer because it stimulates the sensory system effectively (such as by being loud or bright or a colour the observer is very sensitive to). Sensory exploitation might also work better with stronger, more intense signals, even if this results in a less effective resemblance to another object because this more intensely stimulates the sensory system of the animal being duped. For example, an anglerfish might wave its lure much more quickly back and forth than any prey animal would normally swim because this rapid movement is more likely to attract attention. If, on the other hand, it was directly mimicking an actual prey item, it should copy the swimming behaviour closely. All things being equal, mimicry should become more accurate with time because this promotes misclassification. A spider mimicking a flower to lure prey should look more like the flower colour and structure over time because it is less likely to be seen as a fake. In contrast, a spider exploiting a preference in pollinators for, say, yellow objects might simply evolve to be a brighter, more intense yellow, even if this hue goes beyond any yellow flowers that actually exist.

Some scientists, like Schaefer and Ruxton, have argued that sensory exploitation may act as an evolutionary precursor to mimicry because exploitation often involves species or traits taking on the appearance of other salient objects in the environment.[36] In the case of the anglerfish, one might suppose that over evolutionary time the lure started off as an effective means of attracting attention but then became increasingly similar to a real prey item in shape and size if this was more likely to fool its victim. In fact, mimicry and sensory exploitation could even work simultaneously. For instance, an anglerfish lure may resemble the shape and size of a specific object that it mimics, as this is good for luring victims hunting for certain species of prey, yet at the same time the lure may be moved very rapidly, with enhanced movement being better at drawing attention.

When genuine mimicry exists, high accuracy could actually carry a cost. Were the anglerfish lure to very closely mimic one species of prey, then this may be attractive to only a subset of potential targets, reducing the dietary range and possible prey items the anglerfish could lure. The bolas spiders, for example, may be very good at luring one or two specific species of moth, but they will be unable to attract many others. That said, this level of sophistication would be essential for mimicry to work when targeting a response that is very specific to one species. Without mimicking the pheromones of a given species closely, no species would be attracted at all, so in this case specialism is essential. Whether a general form of deception or a more specialist one evolves will partly depend on how precisely tuned the responses of potential victims are to a type of stimulus (like a specific smell or sound). Overall then, deception has several evolutionary routes, and it can be a complicated task to know exactly how it operates, but it can tell us much about how evolution and species interactions work.

We have spent much of the first three chapters discussing how animals and plants use deception to obtain food and prey, and the mechanisms by which deception can work and evolve. Yet deception exists in many other contexts. In Chapter 4, we will begin to look at how animals and plants fight back against risks and threats by using deception for protection, starting with camouflage.

4

DISRUPTION AND DAZZLE

· · · · · · · · · · · ·

In nature, the risk of being eaten is one of the most significant pressures faced by many organisms. Evolution is largely about passing on your genes to the next generation, but that's no good if you die before you can reproduce, and predators are everywhere. We should not be surprised then that animals have numerous and varied ways to prevent being eaten. Many species, such as bison and zebra, live in protective herds, or possess toxic chemicals and unpleasant tastes, as do many caterpillars. Although animals have evolved remarkably sophisticated and varied ways to capture prey, the evolution of protective strategies against predators is often thought to be under even greater selective pressure. One frequently discussed reason is that if a predator fails to acquire its prey then it foregoes a meal, but if the prey animal loses then it dies. This is the so-called 'life–dinner principle', proposed by Richard Dawkins and John Krebs back in the late 1970s.[1] It means that the relationship between predator and prey is asymmetric, and we might expect prey to have a greater range of specialist defensive adaptations than predators have corresponding capture techniques. Whether or not this is true is hard to say for sure, yet there's no doubt that prey have many tactics to deceive predators in order to survive. In Chapters 4–6 we will discuss many of these, how they work and evolve, and how they mislead predators.

Perhaps the most widespread of all animal defences is camouflage. Here, the colours and patterns of animals (and some plants) either generally blend in with the background environment or resemble some specific object around them (like a twig or bird dropping). In Chapter 1 we introduced deception as involving one individual or group exploiting the communication system of others to their own advantage by creating false, exaggerated, or misleading information. Some scientists, including me, have, strictly speaking, argued that camouflage is not consistent with this description of deception because camouflage does not exploit a communication system.[2] Unlike the pheromones used by bolas spiders to exploit the mating behaviours of moths, most types of camouflage essentially involve an animal simply not being seen. In effect, a camouflaged moth is trying to prevent any signal from existing about its presence. However, not all biologists agree with this argument—instead they take the view that communication and camouflage simply involve changing the behaviour of another animal (such as a predator).[3] In this instance, the behaviour of the predator has changed, at a cost, because it no longer detects a camouflaged prey item that it would have otherwise attacked. Here we need not worry too much about semantics, but instead we can loosely consider camouflage as a form of deception because it involves some sort of trickery or manipulation of others. In addition, there are a number of parallels with how camouflage works and other forms of deception that are valuable to explore.

Camouflage varies greatly in complexity, ranging from the relatively simple white fur of snowshoe hares in winter through to sophisticated mimicry of dead leaves by leaf-tailed geckos, or even being almost entirely transparent, as are some fish (Figure 19). As we will discover, the study of camouflage has (alongside some forms of mimicry) been intricately linked with the ideas of the earliest evolutionary biologists more than most other areas of deception, and provided some of the strongest initial and subsequent evidence for evolution. Not only that,

FIG. 19. The remarkable camouflage of some animals, including the Mozambique nightjar (*Caprimulgus fossii*, top left), a leaf-mimicking bush cricket (top right), a juvenile horned ghost crab (*Ocypode ceratophthalma*, bottom left), and a grasshopper (bottom right), all blending in with the background.

Images Martin Stevens

but studies of natural camouflage have often had close ties with areas of human application, including in art and fashion and for military defences.[4] Much work has also been aimed at understanding how camouflage defeats the visual systems of predatory animals, and this can reveal much about how deception works.

Darwin is rightly revered for his theories of evolution by natural and sexual selection, in addition to a host of clever (and usually correct) ideas about nature and how it works. He spent much time studying and discussing the morphology of animals too, from barnacles to peacocks. Yet he rarely discussed camouflage in any detail, despite the fact that it provided some of the best early examples of evolution. Fortunately, however, camouflage was a key topic of interest for many other early evolutionary biologists and naturalists, most notably Wallace. Wallace was one of the great Victorian naturalist-explorers, spending many years travelling in South America and South East Asia collecting specimens to study and to send back to Britain to sell in order to finance his travels. The camouflage of some of the remarkable animals he observed, such as the extremely specialist leaf-mimicking *Kallima* butterflies from South and East Asia (to which we will return shortly), along with other impressive adaptations in nature, encouraged him to consider the origins of species. Indeed, he wrote many essays on animal coloration and devoted considerable space in his books to discussions of camouflage. In his classic 1889 book, *Darwinism*, Wallace said: 'The fact that first strikes us in our examination of the colours of animals as a whole, is the close relation that exists between these colours and the general environment...serving to conceal the herbivorous species from their enemies.'[5]

Wallace was not alone in his fascination with camouflage and its value in illustrating evolution and natural selection. Even before him, Charles Darwin's grandfather, Erasmus, wrote about it in the context of evolution through his poetry and medical books.[6] Several other prominent naturalists contemporary with Wallace investigated camouflage and how it might work too. One of these was the distinguished Oxford academic Edward Bagnall Poulton. Not only was Poulton a staunch defender of Darwinism and a highly regarded scientist, he also conceived several key theories about the types of camouflage that may exist, and conducted experiments to investigate them.[7] In some respects, he was stimulated by the writings of Wallace on protective coloration, and published his own classic book in 1890 called *The Colours of Animals*,[8] which focussed on insects, where he outlined many theories relating to camouflage and other forms of protective coloration in nature. We will return to these ideas shortly.

Camouflage was ultimately to provide the first clear evidence of evolution, observable in time spans of human lives: the peppered moth (*Biston betularia*) and industrial melanism. Few moths have been studied in quite so much detail as this small, unassuming species. As a textbook example of evolution, and one of the most celebrated examples of camouflage and predation, the peppered moth is worth exploring in detail.

The peppered moth is a relatively common species in the UK and in much of the temperate world, flying at night and resting during the day. In a nutshell, the story goes as follows:[9] different looking morphs (types) of the peppered moth survive differentially against avian predators in polluted and unpolluted woodland. The *typica* form, which is pale with black specks, is camouflaged from birds in unpolluted woodland against pale or lichen-covered trees, whereas the melanic (dark) form, *carbonaria*, is better concealed in polluted woodland, where lichen has been killed and soot has darkened tree bark (Figure 20). In the mid to late eighteenth century, only one form of the peppered moth was known, the *typica* form. Then, in 1848, an amateur lepidopterist found a melanic form, *carbonaria*. The rise of the melanic form was dramatic, so that by the start of the twentieth century it was more numerous in frequency

FIG. 20. The peppered moth (*Biston betularia*). The light-coloured typical form (left) is camouflaged against lichen-covered tree bark, whereas the dark melanic form (right) is well hidden against soot-covered trees in polluted areas.

Images © ampics/123RF

than the typical, pale form, even comprising 90 per cent of the population in some areas. Other similar changes were observed in different moth species at the same time, and all these corresponded with the marked increase in soot and pollution in post-industrial Britain. Then, following anti-pollution legislation in the 1950s and 1960s (notably the Clean Air Act of 1956), the situation reversed and the melanic form declined and the typical form recovered. These patterns were paralleled in continental Europe and North America,[10] showing close matches in the changes in morph frequency with pollution levels.

Iconic work, showing that changes in peppered moths were driven by differential camouflage and bird predation, was performed by Bernard Kettlewell from the early 1950s.[11] Kettlewell, a skilled lepidopterist, was recruited by the famous Oxford geneticist E.B. Ford to conduct experiments to test the theory that the rise in the melanic form was largely because it was better camouflaged from birds in more polluted areas. Kettlewell's most famous experiments comprised several components, including testing whether birds were the drivers of predation on moths. This would seem obvious today, with numerous experiments demonstrating avian predation on insects, but it was far from clear-cut in Kettlewell's time. First, he assessed whether the melanic moths were easier to see against pale backgrounds than the typical form, but more camouflaged against dark backgrounds. To do this he came up with a rough scoring system to determine relative conspicuousness. In addition, he conducted preliminary experiments with captive great tits confirming his assessments, showing that the birds were more likely to find and consume more conspicuous moths.

Next, Kettlewell conducted field studies where he marked and released large numbers of moths first into polluted woodland near Birmingham, and later into unpolluted woodland in Dorset. As part of these studies he released live moths in the morning and checked them during the day to see if they were still present. He observed that the melanic moths were harder to see when released in the polluted woods, and that more conspicuous moths were more likely to be eaten. He even directly observed a number of predation events by birds. Kettlewell also released large numbers of marked moths and subsequently recaptured as many individuals as possible, using lights and moth traps. The logic was simple: melanic forms should be better hidden in polluted woodland and so more should survive and be recaptured, whereas the opposite should occur in the clean woodland. This is exactly what he found. In the polluted woodland he recaptured 27.5 per cent of the melanic moths but just 13 per cent of the typical ones, whereas in the clean woodland he recaptured 6 per cent of the melanic moths and nearly 13 per cent of the pale forms (which were also much less conspicuous to Kettlewell). When he repeated the Birmingham experiments, 52 per cent of the melanic moths but only 25 per cent of the typical ones were recaptured.

Finally, to provide further evidence that his findings were driven by bird predation, Kettlewell recruited Niko Tinbergen (who won a Nobel Prize for his work on ethology) to film birds capturing moths and to convince sceptics that birds really did eat them. The filming took place during the second Birmingham experiment, in which Kettlewell released equal numbers of the melanic and typical form at the same time on to tree trunks in view of a hide. From this, Tinbergen filmed fifty-eight predation attempts, which were primarily on moths of the *typica* form, showing that the birds were eating the moths and that they were more likely to overlook the camouflaged ones. In the Dorset experiment, Tinbergen again recorded predation on moths released on to trunks but in the opposite direction; this time, as expected, the melanic moths were more often eaten.

The peppered moth story and industrial melanism was rightly adopted as one of the most celebrated examples of evolution in nature. Yet, in the late 1990s it started to come under attack. The main reason relates to an agenda by anti-evolution creationists who were seeking ways to attack evolution and the teaching of it. They picked up on discussions at the time between biologists about the details of Kettlewell's original work and some of the unresolved issues surrounding the example. Some creationists misrepresented the scientists involved to create the impression that there were doubts about the conventional explanation. In actual fact, the vast majority of scientists and articles did not doubt the overall validity of Kettlewell's findings, or that the main driving force behind the peppered moth case was differential camouflage and bird predation. Around the same time, however, was the publication of a popularized account of Kettlewell and Ford's work on the peppered moth, putting forward the suggestion that Kettlewell committed fraud by fabricating his data when initial experiments were not going to plan, and that other Oxford academics (such as Ford) were complicit in some sort of cover-up. Yet, as countless experts have since pointed out, the claims lack any foundation.[12] Sadly, however, the damage had been done and the validity of the example was being more broadly questioned. It is important to note here that we should not hide from the fact that some criticisms of Kettlewell's work are reasonable—notably that camouflage assessment relied on human judgement, even though avian vision differs from our own. Kettlewell also released the moths in large numbers in his fieldwork, and used a mixture of wild and lab reared moths, which may have affected the results to a degree. None of these or other factors cast doubt on the overall conclusions, however. And as we'll now discuss, the peppered moth story is far better tested and supported today than during Kettlewell's time.

First, although they did not stop the creationists, a spate of reviews and papers went some way to re-establishing the truth behind Kettlewell's work.[12] In addition, research continued on the peppered moth, and many studies retested Kettlewell's original work between

1966 and 1987, showing remarkably consistent findings with each other and with the original research.[9] This greatly enhanced the evidence that bird predation was the primary force in the rise and fall of the melanic form. Mike Majerus, one of those scientists misrepresented in the supposed controversy and an expert on the peppered moth, used cameras capable of imaging in UV light (which birds can see) to show that the typical form should be well camouflaged against lichen to the eyes of a bird.[13] Beyond this, the peppered moth is actually a doubly impressive example of evolution in action because, since Kettlewell's original work, there has been a decline in the melanic morph. Laurence Cook and others from the University of Manchester reported changes in the frequencies of morph types in the Manchester area of Britain in the 1960s, with the typical form becoming relatively more common. This coincided with clean air legislation in Britain in the 1950s and 1960s, such that by the mid-1980s the melanic form was in serious decline; Cook and others calculated that the melanic form was at a disadvantage of about 12 per cent compared to just twenty years earlier. This decline of the melanic form has been very well documented elsewhere too.[14]

Finally, Majerus went to great lengths to set the record straight. I met him a few times in Cambridge and heard him give seminars on the subject; he clearly felt great frustration that such a wonderful example of evolution was being unfairly attacked. He told me about the work he was doing to show the importance of bird predation in driving changes in morph frequencies over time (this time regarding the decline of the melanic form). One of his main studies involved a six-year experiment in Cambridgeshire between 2001 and 2007, during which he released 4,864 moths to measure bird predation. This was perhaps the largest predation experiment ever conducted.

Sadly, Majerus died in 2009 before much of his work could be published. However, it was not in vain. In 2012 a group of scientists who had also spent many years studying the peppered moth were able to use and analyse Majerus' data and publish them in a scientific report.[15] The main part of his experiment involved releasing moths of melanic and typical forms in approximately the same frequencies as they naturally occurred, and measuring bird predation on each. Majerus released a single moth into netting sleeves placed on trees around different branches and allowed them to settle naturally during the night. In the morning, before dawn, the netting was removed and he monitored avian predation. Majerus directly observed 26 per cent of the moths he recorded as eaten being attacked by birds, and the study clearly showed that more melanics than pale forms were eaten overall. All in all, there was significant selection against the melanic form in the unpolluted site, with melanics having a survival probability of between 84–97 per cent of that of the typical form (the average was 91 per cent). A statement by Laurence Cook and colleagues in their paper's conclusions about Majerus' work should leave everyone in no doubt about the validity of the peppered moth example: 'with this new evidence added to the existing data, it is virtually impossible to escape the previously accepted conclusion that visual predation by birds is the major cause of rapid changes in frequency of melanic peppered moths'.[16]

Camouflage and the peppered moth provide a wonderful example of evolution. However, there's a rich complexity underpinning how camouflage actually works, and for that matter why it works. Camouflage has been a primary focus of my research for over a decade, and one of the main questions my colleagues and I have asked is how exactly it works in defeating predator vision. In some ways this might seem obvious: camouflage involves simply resembling another object or background in the environment. But this answer fails to appreciate the many intricacies and sophistication involved in how camouflage achieves its goal, and that there are many methods of concealment that could work in practice. Let's begin with probably the most widespread and conceptually straightforward type of camouflage found in nature (and the type of concealment used by the peppered moth), a strategy called background matching. Here, an object (e.g. an animal) resembles the general colour and pattern of the background environment on which it is found. This means that if you're a moth resting on the trunk of a tree, you should match the general colour and pattern of the bark. Quite simply, provided the match is good enough, a predator searching for a hidden moth would fail to detect one in front of it. Background matching was the type of camouflage most often considered by those early evolutionary biologists like Wallace, but as with most early studies of animal coloration, it was often viewed from a human visual perspective.

As Wallace noted, background matching means that related populations of animals found on backgrounds that look different should develop or evolve divergent appearances, in accordance with coming to resemble their respective environments.[17] For example, fish that live on orange coral should over time become orange, whereas those that live on blue coral should become blue. Conversely, unrelated groups of animals living in the same habitat may often converge in appearance, just as many desert animals, from desert foxes to small lizards, are yellow-brown in colour. These expectations have been well supported in various studies, but perhaps most effectively in recent research on mice and lizards. For example, Hopi Hoekstra and colleagues from Harvard University have shown that individual American pocket and oldfield mice are light brown in colour when they live on pale substrates, such as sand dunes, but dark in colour where the ground is very dark, such as old lava flows. The basis for these differences in coloration stems from changes in a limited number of interacting genes as the populations diverge.[18] Very similar findings, again involving changes to a limited number of genes, have been found in several species of lizard living in light or dark substrate areas of New Mexico.[19] In the case of the mice, experiments using fake plasticine models resembling light- or dark-coloured individuals show that mice are less likely to be attacked by predators (such as birds) when found on their associated backgrounds than on mismatched backgrounds.[20] In fact the selection is very strong, demonstrating why we so commonly see clear links in colour between animals and the environments in which they live for camouflage.

In other species, associations between animal appearance and the coloration of their environment is driven more by changes that occur during individual development. One of the main study species in my research group is the shore crab (*Carcinus maenas*), probably the most common crab species in the UK and much of Europe (and sadly a rampantly invasive species all over the world now too). It is a highly adaptable animal found in many habitats from mudflats through to mussel beds and rock pools (Figure 21).

Shore crabs are interesting because individuals are not only very well camouflaged but are also strikingly variable in appearance from one crab to another, both in coloration and pattern; this is especially true for the juveniles. Quite why they are so variable, especially the young ones, is something of a mystery we are yet to fully resolve (Figure 22). What we do know, based both on studies undertaken in Scotland by Peter Todd of the National University of Singapore and colleagues, and work in my lab in the south-west of the UK (Cornwall), is that the coloration of individual crabs varies with habitat type, almost certainly to provide camouflage to match their respective backgrounds.[21] Unlike the oldfield mice mentioned earlier, genetic differences between the crabs are unlikely to underlie most of these differences because the study sites are close together geographically and the crabs reproduce by spawning into the open ocean, meaning that the gene pool across all these habitats gets strongly mixed up at the planktonic larval stage. Instead, most changes must occur after juvenile crabs settle in a particular location. Again, we are still figuring out exactly what's going on, but crabs seem to be able

FIG. 21. Camouflage of the shore crab (*Carcinus maenas*) against different rock pool backgrounds. Shore crabs are highly variable in colours and patterns and this enables individuals to blend in with and match a wide range of substrate types.

Images Martin Stevens

to change colour over a period of hours and days as they land on a particular background, and then change appearance more dramatically over a period of weeks and months as they moult and change the colour patterns on their exoskeletons. Over time, they become better at matching their environment. The process is called 'developmental plasticity', and it is probably guided by input from the crabs' visual systems as they look at the surrounding environment. This fantastic

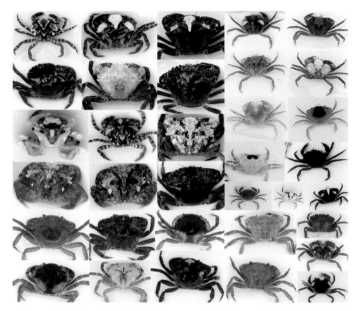

FIG. 22. The impressively variable appearance of shore crabs (*Carcinus maenas*). All individuals here are of the same species and found on the same small beach in Cornwall, UK. Crabs here are of various ages and sizes, but adults and sub-adults are mainly on the left and centre, and juveniles are mainly on the right.

Images Martin Stevens

ability means that crabs can develop different body colours and patterns to achieve successful camouflage on a wide range of background types even without undergoing genetic divergence among populations.

That animals evolve and develop camouflage patterns to match their backgrounds, both among and within species, is uncontroversial (though there's much we don't know about how it works). However, what evidence exists that background matching actually prevents detection from predators? Some of the earliest proof came back in 1977 when Alexandra Pietrewicz and Alan Kamil from the University of Massachusetts showed how the camouflage of moths deceives avian predators.[22] Pietrewicz and Kamil trained six hand-raised blue jays (*Cyanocitta cristata*) to respond differently to the presence or absence of moths in projected slides. Jays were trained to peck at a button, a so-called 'stimulus key', if they detected a moth in the image. If the jays got this right they were given a reward in the form of a piece of mealworm. If the birds responded incorrectly, for example by pecking the stimulus key when no moth was present, they were punished with a one-minute time delay before the next slide was shown. The slides were of *Catocala* moths, a group of moths that are well camouflaged against tree bark. The scientists took photographs of moths pinned to trees or logs, as well as corresponding photos of the bark without moths, to generate positive (with moth) and negative (no moth) slides. The birds clearly found it easier to detect moths when they were placed against backgrounds that they did not closely resemble. The results of this study might seem somewhat predictable now, but what made it so clever was the way that Pietrewicz and Kamil's methods and use of technology allowed them to demonstrate that camouflage does work against non-human animals.

I mentioned earlier that many camouflaged species have individuals with varied appearance, sometimes with multiple morphs found in the same location (so-called 'polymorphic' species). There exist several explanations for this phenomenon, but it's often thought to be another way that camouflage deceives predators. Detection of camouflaged prey can sometimes depend on the learning and cognitive processes of the predator. Imagine being in a supermarket and searching for your favourite chocolate bar among a host of competing products. Your search is made easier if you remember what the packaging on the bar looks like because your brain can focus attention on items that match this appearance, while filtering out things that deviate from it. As a result, you are much more efficient in your search. But there's also a cost. By focussing on one specific appearance, you become worse at finding other non-matching items. So you might, for example, miss out on a new, even better, chocolate bar further down the aisle. This concept, widely called a 'search image', has been around for a long time. In 1890,

Poulton suggested that searching for several different prey items at once is more challenging for a predator than looking for one thing at a time.[8] The general idea is that, through recent experience, the visual system and brain become primed to focus attention on finding prey of a given appearance. Nonetheless, despite substantial research and scientific discussion, the idea of a search image is not without controversy. This is because it has proved notoriously hard to demonstrate the effect of a genuine search image, as opposed to other factors that could equally explain improved foraging performance, such as predators searching their environment faster (increased search rate) or becoming better at knowing where in the environment to find prey (for example learning that a moth type is more common on the outer branches of a tree).

Search image theory has some genuinely interesting implications for the evolution of camouflage and prey coloration, and for selection processes in general. In particular, one prediction is that it should lead to prey species that are polymorphic. In nature, less-common prey types will, all things being equal, be encountered less frequently than common prey forms, meaning that predators have less opportunity to form search images for rarer morphs. Consequently, rare morphs have a lower risk of predation and begin to increase in numbers. This is the case until they become more common, at which point predators should switch and form search images for them instead. The effect is that prey individuals are less likely to be eaten if individuals come in a variety of forms because predators may focus on the colour patterns of one morph at a time and overlook others. It also means that prey populations should fluctuate or cycle in the relative proportions of morphs over time. This process is often referred to as 'negative frequency-dependent selection', whereby less-common types are at a selective advantage.

So far, evidence for search images and their effect in the natural world is limited. Perhaps this is unsurprising, because the concepts do not easily lend themselves to study in the wild. However, observationally, many camouflaged prey, including moths, grasshoppers, and crabs, do exist in multiple morphs occurring in the same geographical locations, and several classic studies have shown that morph types do fluctuate over time, consistent with negative frequency-dependent selection. The best evidence, however, for search images working in predators, and the effect on prey appearance, again arises from clever experiments with captive blue jays.

Two years after their paper presenting hidden moths on projection slides to jays, Pietrewicz and Kamil published another study using similar methods to test the search image concept.[23] They presented jays with sequences of slides of hidden moths and looked at detection times over successive presentations. To some birds, they presented runs whereby the bird saw only one moth species (moths of the same appearance), whereas other birds were given sequences that contained mixes of two moth species (which looked different). The key prediction was that the detection abilities of jays should improve only for birds that saw just one species. The logic behind this was that in the mixed presentations the occurrence of moths with different appearances would prevent the birds from being able to focus their attention on one species only, thus hindering search image formation. This was exactly what Pietrewicz and Kamil found; performance over trials improved substantially when the birds saw only one species, yet little improvement occurred when birds encountered two species.

It was not until the end of the 1990s and early 2000s that significant progress on the issue of prey polymorphism and frequency-dependent selection was made. A series of ingenious studies by Alan Bond and Alan Kamil (now at the University of Nebraska) presented blue jays with 'virtual moths' on computer screens to test the interplay between search behaviour and the frequency of different moth types[24] (Figure 23). Initially, they presented birds with moths of a limited

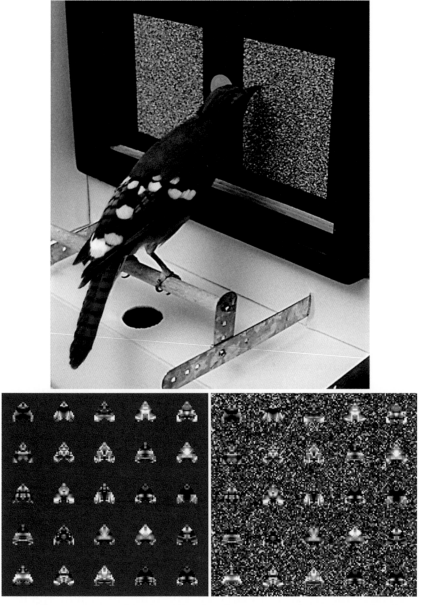

FIG. 23. The experiments of Bond and Kamil used to test prey polymorphism and predator search images. Top shows a blue jay searching for a hidden artificial moth in the background. The bottom images show examples of the moths that evolved, and their diversity in appearance (shown on a uniform grey background, and against the background used in the experiments).

Images Alan Bond

number of fixed pattern types, and showed that these fluctuated in frequency over time, as would be expected under negative frequency-dependent selection. In later experiments, the authors gave the moths a computer genetic algorithm, a bit like a computer-based 'genome' that encoded each moth's appearance, such that they could evolve new patterns through repeated encounters with the birds. As predicted, moth populations evolved a wide range of different pattern types, and in turn the jays often failed to find rare or novel morphs, consistent with search image theory.

To successfully deceive predators, matching the colour and pattern of the background is clearly crucial. But we should not forget the role of behaviour too. Yet again, Wallace was among the first to appreciate this. He noted that the leaf-mimicking *Kallima* butterflies he encountered on his travels only landed in places where their disguise would work. He reported seeing scores of them in Sumatra, and found that they never landed on flowers or green leaves, but instead always settled within bushes or on piles of dead leaves, holding themselves against branches so that the 'tails' of the hindwings touched the plant to look like a stalk.[17] Likewise, Kettlewell pointed out that cryptic animals should rest on backgrounds against which their camouflage is tuned. It's no good evolving wonderful camouflage against the bark of an oak tree if you end up resting on the green leaves. As part of his work on the peppered moth, Kettlewell conducted a very simple experiment whereby he lined the inside of a large cider barrel with alternating black and white stripes (controlling for the overall surface area of each).[25] Each evening he released three light and three dark peppered moths into the barrel and recorded their resting positions in the morning. At the end of the experiment Kettlewell found that for each morph type approximately twice as many moths chose the correct background (i.e. white for light moths, dark for melanic moths) than chose the wrong background.

Subsequent work in the 1960s by Ted Sargent from the University of Massachusetts supported Kettlewell's findings. Sargent used a very similar approach, with an experimental box painted in different shades of grey, into which he released moths from eight species. Darker species of moth tended to rest on darker backgrounds, and pale species on lighter backgrounds.[26] Like Kettlewell, Sargent also found that individual forms of a species could differ in background choice. Thus, many species and morphs of moth seem capable of choosing the appropriate background for their camouflage. Kettlewell suggested that the moths somehow compared the contrast between the body scales surrounding their eyes with the background to guide this behaviour. However, when Sargent painted over the scales of moths to see if it affected their background choice it had no effect, indicating that the choice is likely to be fixed in individuals of a species (innate), or controlled by some other mechanism, such as assessing the texture of the substrate.[27]

The above principles do not just apply to moths and trees. Any behaviour that improves the camouflage match of an individual will place it at an advantage. Many ground-nesting birds such as Japanese quail (Coturnix japonica) have well-camouflaged eggs to avoid predation. The eggs of different individuals also vary from one mother to another. Recently, George Lovell and colleagues from St Andrews University asked whether mother quail choose backgrounds on which to nest that improve the camouflage of their own individual eggs (Figure 24).[28] They gave quail a choice of different substrates on which to lay that differed in how bright or dark they were, and found that females chose backgrounds that were more similar to their own specific egg colours. That is, individuals with light egg colours chose light backgrounds, and those with dark eggs chose dark backgrounds. This implies that in the wild, ground-nesting birds might choose patches of ground on which to lay their eggs that offer the best camouflage for their own egg markings.

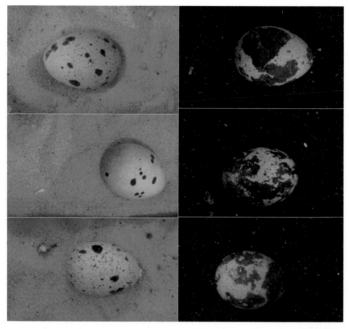

FIG. 24. Japanese quail (*Coturnix japonica*) egg patterns that vary from one female to the next. When given a choice, females with lighter eggs prefer to lay on light substrates, whereas those with dark patterns choose dark substrates, thus improving their individual egg camouflage through behaviour.

Images George Lovell/Keri Langridge

Choosing the correct background may still not always be enough to optimize concealment. Another key factor is likely to be the way that an animal orientates its body. In his peppered moth experiments, Kettlewell suggested that, rather than searching the woodland for the best sites to rest, moths instead come to rest on a tree and then locally shift their location to find the best position with regard to their coloration (morph type). Let's consider another moth camouflaged against a tree trunk. Bark frequently has ridges running down it, making lines and small gullies extending vertically down the trunk. The markings of many moths resemble these patterns, but this also means that the moth must rest in a position such that the lines on the tree and their own body coincide. It's no good sitting on a tree with your markings orientated to be horizontal if the tree patterns are mainly vertical. Somehow, the moth needs to orientate itself in the right position to ensure its camouflage is effective. This was suggested by Sargent,[29] and recent experiments have shown that some moths do just this. Changku Kang and colleagues at Seoul National University released moths on to trees and analysed what they did after landing. In many cases, individuals changed from their initial landing position and orientation to a new one. What's more, in these new positions the moths were harder to detect than in their original positions, both to humans and to models testing how the moths look to bird vision. Subsequently, the authors also showed that whether or not the moths reposition themselves depends on how well hidden they were initially; moths that first landed in a location that provided good camouflage did not reposition themselves.[30]

These various tricks show how nuanced and sophisticated camouflage can be to evade the searching eyes of predators. Yet many animals have another method, one that enables them to be well hidden on a large number of background types in a relatively short space of time: colour change. The kings of colour change for concealment are cephalopods (cuttlefish, octopus, and squids).[31] They are eaten by many predator groups, including fish, diving birds, and marine mammals, and live in a wide array of habitats, from coral reefs and kelp forests to sandy environments. Cephalopods show a staggering ability not just to change colour rapidly (in a few seconds), but to also select a wide range of pattern types depending on the background they are viewed against (Figure 25). They have been extensively studied, in particular by Roger Hanlon and his team at Woods Hole Oceanographic Institute near Boston.

Cephalopods' and other animals' abilities to change their appearance derive from special cells in the body, in particular chromatophores.

FIG. 25. The disguises of the Australian giant cuttlefish (*Sepia apama*). Top left: undisguised. Bottom left: resembling the colour and pattern of the background. The right-hand images show it blending in and mimicking both the colour and 3D structure of objects in the environment.

Images Roger Hanlon

Chromatophores contain little packages of pigment used to change the body colour and darkness, and in cephalopods have muscles attached to them that are directly under the control of nerve cells (explaining their rapid action). Impulses from these nerve cells cause the chromatophores to expand or contract, spreading out or concentrating the pigment, and changing the colour and pattern of the body surface. This enables them to change the contrast, size, shape, and distribution of markings on their body to resemble a range of substrates. Many cephalopods can even change the three-dimensional nature of their skin in order to match the structure or texture of the surface they are trying to resemble.

Cuttlefish (especially the European cuttlefish, *Sepia officinalis*) in particular have been a key group in which to study how camouflage works. Hanlon's group and several other research teams have conducted many experiments in the laboratory (as well as observations while diving), often involving putting cuttlefish on controlled backgrounds of different appearance and testing the patterns the animals produce.[32] They have found that cuttlefish are clearly capable of matching many background types, with the patterns produced ranging from relatively uniform appearances, to small 'mottled' markings, and on to larger and high-contrast markings. For example, relatively larger and high-contrast body patterns tend to arise on substrates with sufficiently large patterns (compared to the cuttlefish's body size), and in particular when in the presence of patterns with clearly defined edges. On low-contrast substrate patterns, cuttlefish generally adopt more uniform markings.

The ability to change colour for camouflage (among other functions, such as for communication with rivals or mates) might seem a special talent, indeed Wallace thought it 'rare and quite exceptional',[33] but it is much more widespread among animals than is often realized. Beyond cephalopods, colour change over varied timescales is found in crabs and shrimp, various fish, spiders, caterpillars, and perhaps most famously in chameleons, to mention just a few groups. The changes can often be rapid (seconds or less), caused by direct signals from the nervous system, or they can be slower (minutes, hours, or days), such as when they are controlled by hormones, or even seasonal in line with snow cover as in some mammals and birds. Some forms of colour change can take weeks if they result from dietary changes or occur during development, as determined by Poulton, who conducted some basic experiments in the 1880s and 1890s with caterpillars. Those experiments and more recent work have shown that both background coloration and diet can play a role in driving colour change and camouflage in caterpillars.[34]

As yet, it's not entirely clear why some animals have the ability to change colour whereas others do not. One likely advantage is that colour change enables animals to cope with varying backgrounds and unpredictable environments. For example, in my lab we have spent the last few years conducting studies investigating colour change in animals living in intertidal habitats such as rock pools. It is clear that many species, including various fish and some crabs (including the shore crabs we encountered earlier),[35] can change colour to improve their camouflage. Some rock pool gobies can change both their general brightness and their coloration when placed on different backgrounds.[36] They become redder when placed on a red substrate, with most of their colour change taking just one minute to complete. For the gobies, this colour change is likely to be a matter of life or death. The rock pool environment is challenging because the substrates (rocks, sand, gravel, seaweed, and so on) look very different, and waves and tides can push them against these during the day. What's more, at high tide they are almost certainly eaten by larger fish and other marine predators, whereas at low tide they risk bird attacks. So being able to change colour quickly is important in avoiding being eaten in such a testing environment.

We have spent most of this chapter so far discussing background matching. However, resembling the general appearance of the environment might not always be as effective as we might first imagine. For one thing, this process often leaves an unbroken outline of the prey's body shape that can be detected by a predator. For instance, the characteristic shape of a moth or a bird egg may be visible even if the object matches the environment effectively. Furthermore, predators can often use characteristic body features such as the edges of wings, legs, eyes, and various other cues to detect a hidden prey animal. To be hidden effectively therefore requires destruction of these tell-tale features. How can this be achieved? One major avenue involves a type of camouflage known as disruptive coloration. The concept of disruptive camouflage, and related theories, has a long history, being initially proposed by two individuals who remain key sources for many ideas in animal camouflage and coloration, although they came from very different backgrounds. One of these, Poulton, we have already encountered. The other was the American artist Abbott H. Thayer. Both were early proponents of Darwinian evolution, and both were advocates of camouflage as being an important example of the power of evolution and adaptation. Almost simultaneously, towards the end of the nineteenth century, they each came up with ideas for how to break up body shapes. However, it was some time before their ideas became accepted or even scientifically tested.

Thayer in particular went into great detail about disruptive coloration (or 'ruptive' coloration, as he called it). In 1909 he, along with his son Gerald, published a substantial and influential book, *Concealing-Coloration in the Animal Kingdom*.[37] In it he outlined thoroughly and at great length his various theories for animal camouflage, and presented concepts and ideas that would, almost one hundred years later, spur resurgence in camouflage work and outline many of the key principles of how camouflage defeats predator vision. Disruptive coloration was one of the most important of these concepts, with Thayer describing and drawing how high-contrast bold patches of colour could destroy the shape and appearance of an animal's body (Figure 26). Poulton described much the same thing with regard to some caterpillar markings in 1890,[8] although his ideas on this subject were clearly far less developed. Thayer also had a role in stimulating the formation of British, American, and French camouflage units and ideas for camouflage on humans and machinery.[38] Indeed, Thayer was convinced that his knowledge and theories could save many lives during the First World War, although his pleas to the relevant military parties were not always well received and he was frequently ignored. He was also pioneering in the practical way in which he presented his theories. He would frequently set up interactive displays for colleagues and the public, or ask subjects to look

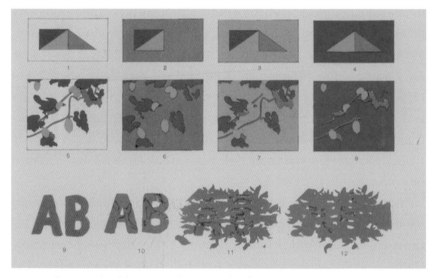

FIG. 26. The principle of disruptive coloration, whereby strong patches of colour or contrast, some of which blend into the background, can be used to break up and destroy the outline of the body. From Thayer (1909).

The Bodleian Libraries, The University of Oxford. Thayer, G. and Thayer A. 1909. Concealing coloration in the animal kingdom. Macmillan Co. New York. Plate 5 (after p 76)

for hidden models of things such as birds in trees to convince them of his concepts. Thayer met both Wallace and Poulton in the UK when he was giving some of his demonstrations. Poulton in particular was impressed (unsurprisingly, given that he had independently had much the same ideas), and the two of them remained friends thereafter.

Unfortunately, Thayer's and Poulton's theories were largely lost from science for some years, and Thayer was more often the source of ridicule by biologists instead. In truth, he did not help himself with the way he presented his ideas. In his 1909 book, he stated how protective coloration '...has naturally been considered part of the zoologists' province. But it properly belongs to the realm of pictorial art, and can be interpreted only by painters.' The book, he stated, 'presents not theories but revelations'.[37] Such statements and an arrogant tone won him no favours, and many zoologists lost patience and his work became largely forgotten. He also took his ideas to ludicrous extremes, arguing that *all* animal colours were used in concealment, even the pink coloration of flamingos at sunset and the gaudy tail of the peacock in the forest. Such ideas were clearly wrong, and became used in undergraduate biology lectures as examples of 'just-so stories' that sounded nice but had no evidence to support them. The famous palaeontologist and science writer Stephen J. Gould even referred to Thayer's example of camouflaged flamingos as an example of 'illogic and unreason'.[39]

Thayer's cause was also not helped by a series of high-profile spats with Theodore Roosevelt, who at the time had just finished his term as US President. Roosevelt was an avid natural historian and was used to seeing animals in nature through travelling and safaris. He poured scorn on Thayer and his ideas. He (wrongly) questioned the role of natural selection in producing animal coloration, and (rightly) questioned some of Thayer's more extreme arguments about how all animals were camouflaged. Roosevelt took issue even with Thayer's demonstrations, stating in a letter that: 'You might be illustrating something in optics, but you would not be illustrating anything that had any bearing whatsoever on the part played by coloration of the animal in actual life.'[40] Here, Roosevelt partly missed the point of Thayer's arguments. While Thayer clearly selected the backgrounds against which he displayed animals carefully, this was to demonstrate a proof of concept, and to illustrate just how his theories of camouflage could work and how effective they can be. In any case, Roosevelt was far from the only person to heavily criticize Thayer and his ideas, and much debate and furore surrounded his work at the time. Sadly, Thayer, who seemed to suffer from a form of bipolar disorder or depression, was very down at the time of his death in 1921. A lack of acceptance of his theories and ideas of natural and military camouflage seems partly responsible.

As it happens, it was a zoologist, Hugh Cott, who formalized many of the ideas of Poulton and Thayer, outlining and extending those theories in a clear and more scientific manner in a landmark book on animal coloration in 1940.[41] Like Thayer, Cott was also involved with helping to advise on military camouflage and was also an excellent artist and illustrator, and an early pioneer of the use of photography to study animal coloration.[42] In fact, he had greater success and influence than Thayer in guiding military camouflage (at least in the UK) and training others about this concept, no doubt partly due to his military service at home and abroad, particularly during the Second World War. Cott, who among other positions was curator of the Zoology Museum at the University of Cambridge, also travelled extensively in Africa, South America, and the Middle East, studying and drawing many animals in the wild. He laid down his ideas about camouflage, including concepts such as disruptive coloration, in such a clear and powerful way that his book is still a major influence on the field. Perhaps ironically, it may have been the clarity of his arguments that meant work on camouflage was largely halted after his book. As a consequence, ideas such as disruptive coloration became textbook examples of camouflage, despite virtually no scientific experiments having been done to support those ideas. This only changed in the last fifteen years or so.

Disruptive coloration became widely accepted post-1940 as a mechanism of camouflage found in many animal groups. But does it really work, and if so how? One of the ways we can test the survival advantage of concealment is to dispense with real prey animals entirely, and instead make our own artificial 'prey'. During my PhD at Bristol University, my supervisor Innes Cuthill and I, along with various colleagues, did just this.[43] We made fake 'moth' prey based on triangular pieces of waterproof paper, printed with specific colours and patterns from digital photographs of tree bark. We then added a dead mealworm (which birds love to eat) to each target and pinned them to trees in woodland near Bristol. The idea was not to mimic any real prey species, but rather to create a stimulus that broadly resembled a natural type of prey that birds would search for, and that was easy to very carefully and precisely change in appearance. By doing this, we could alter aspects of the colour and pattern of different moth types and see how effectively these were hidden from wild predators (Figure 27). By monitoring the targets every few hours over a series of days we could record which ones were detected and eaten the fastest (as judged by the disappearance of the mealworm). One of our main studies compared the 'survival' of targets that simply matched the colour patterns of tree bark with the survival of targets that were predicted to be disruptive. These latter targets possessed patterns also matching tree trunks, but this time with some markings deliberately placed to touch the outline of the body. Cott and Thayer predicted that for disruption to work some markings should extend to the body outline and blend into the colours of the background. In addition, other patterns should contrast strongly

FIG. 27. Testing the survival value of disruptive coloration against avian predators. Top shows artificial paper 'moth' prey. The top left moth has disruptive markings that break up the outline, whereas the adjacent target on the right does not. The moths below have the same disruptive patterns but of high or low visual contrast, with the former being more effective in promoting disruptive camouflage. Below is an example artificial moth pinned to a tree from another experiment.

Top images Martin Stevens and Innes Cuthill; bottom image Martin Stevens

with markings on the body and background. The effect should destroy the appearance of the body outline and shape. It sounds sensible, but does it work? Yes, emphatically. Over the course of twenty-four hours the disruptive targets had a chance of survival of about 70 per cent, compared to just 30–40 per cent for those non-disruptive prey that simply copied the tree bark patterns. What's more, the disruptive targets were especially effective when they had patterns that had high contrast, just as Cott and Thayer predicted. A range of other subsequent experiments by various researchers has confirmed that disruption works over and above simple background matching.

That answered the question of whether disruptive coloration works, but how does it deceive predator vision? Part of the answer seems to stem from the way that visual systems encode information about objects and boundaries in natural scenes. Much information in the natural world, including the boundaries between objects and shapes, occurs with sharp changes in light intensity, or transitions from light to dark. For example, the shape of a dark tree trunk seen against a light sky is easy to pick out because there's a sharp difference in light intensity where the tree ends and the sky begins. Visual systems are very effective at breaking down information about natural scenes into edges and boundaries that mark the transitions between objects, and the way that this is done appears similar across a range of species. This, for example, can include things like encoding the characteristic shape of a moth against a tree, by virtue of the edges where the moth body touches the tree outline. Coming back to disruptive coloration, it had been suggested that by destroying features of the body shape, and preventing the visual system from encoding information about the body edges, disruption can cause a predator to fail to detect that something is present. Shortly after doing our field experiments, Innes and I set about implementing a model of predator vision based on how visual systems might encode edge information in a visual scene.[44] The model used mathematical algorithms that broadly resembled how vertebrate visual systems might encode boundary information to work out where the edges of objects are. We took photographs of the different target types used in our experiments pinned to trees, ran them through the model, and analysed how much of the moth target outline was left intact. The disruptive targets had far less information about their body edges retained by the vision model than did prey with other camouflage types, and consequently the model was less effective at detecting disruptive targets.

The other recent piece of evidence showing how disruptive coloration works was by Richard Webster, Tom Sherratt, and colleagues from Carleton University in Canada.[45] They designed a computer experiment to test how human subjects wearing eye-tracking devices responded when looking for disruptive 'moth' targets against images of tree trunks on computer screens. Thus, the experiment was essentially similar to the wild ones with paper moths and birds, but this time using humans and a computer 'game'. Humans took longer to detect targets with more of their edges broken up by the disruptive markings. Subjects also looked at disruptive targets for longer before attacking them, and were more likely to pass over disruptive targets while searching. So, it seems that breaking up body edges is a very effective way to hide an object.

One of the really nice things about disruptive coloration is the consistency in results between experimental set ups. Humans are not the natural predators for many of the animals that biologists study (though camouflage is often important to us too), yet when we compare results between humans playing computer trials and birds foraging in the wild or in aviaries, the results are remarkably consistent. This tells us that disruptive coloration is not some quirk of one predator group or one prey type, but a broad principle we should expect to find widely in nature. In fact, it has been suggested that it exists in many animals, from birds such as ringed plovers, to insects like carpet moths. As yet, however, a key gap in research is in demonstrating the occurrence and importance of disruption in real species beyond artificial prey studies. That is, while we expect it to be widespread in nature, nobody has studied this in detail.

I have so far neglected to discuss other types of camouflage beyond background matching and disruption, and there are many.[46] There is, however, one method of being hidden that is much more specialist. It works not by preventing detection, but rather by preventing recognition, and is called masquerade. Most of us will have come across this at some time, even if it is through seeing a stick insect in a pet shop or zoo. It involves an animal resembling a specific object in the environment that would not normally be of any interest to the predator, for example a bird dropping, dead leaf, or stone. The key distinction with most other forms of camouflage is that in masquerade, a predator can detect the prey but it is not correctly recognized or is misclassified as the wrong object type. That is, a bird foraging for a nice juicy caterpillar might see something on a leaf that resembles a bird dropping but fails to appreciate what it really is. Thus, masquerade has direct parallels with mimicry because it involves the observer misclassifying the prey animal as something else.

As we know already, Wallace was hugely impressed with the leaf mimicry of *Kallima* butterflies, calling them 'the most wonderful and undoubted case of protective resemblance in a butterfly'.[17] As he described, the underside of these butterflies beautifully mimics dead or decaying leaves in shape, colour, and through having a dark line running along the wing mimicking the midrib of a leaf, as well as smaller leaf veins. Not only that, but many butterflies appear to resemble leaves in various states of decay and, as Wallace put it, 'so closely resembling the various kinds of minute fungi that grow on dead leaves that it is impossible to avoid thinking at first sight that the butterflies themselves have been attacked by real fungi!' (Figure 28). Interestingly, Wallace even noted that within species of *Kallima*, individuals come in a wide variety of appearance. Perhaps this is a mechanism to prevent predators from learning about them and forming search images. We will return to the evolution of *Kallima* butterflies in Chapter 9.

PLATE II.

Kallima butterfly.

Peter Smit del. et lith.   Mintern Bros. Chromo.

FIG. 28. An early illustration from 1892 of leaf mimicry by *Kallima* butterflies, in this instance by a contemporary of Wallace, Frank E. Beddard, who also wrote in depth about animal coloration, especially in his book *Animal Coloration.*[44]

The Bodleian Libraries / The University Oxford. "Animal Coloration. An account of the principal facts and theories relating to the colours and markings of animals" 1892, Swan Sonnenschein, plate 2

It's almost impossible to doubt that masquerade exists or that it must work. Why else would so many insects look like twigs or leaves? But it's still important to conduct experiments to confirm this is true and how it works, because our perceptions can sometimes be misleading, and because experiments are needed to understand how exactly such adaptations work and evolve, and to establish their survival value. Unfortunately, for something that seems so clear to our eyes, masquerade is actually very hard to demonstrate. The key requirement is to show that the predator fails to classify a prey animal correctly rather than simply not seeing it, and this is challenging because we can't just ask a predator why it didn't attack something. It is only in the past few years that experimental evidence supporting the idea of masquerade, and how we expect it to prevent attack, has been obtained. Research at the University of Glasgow by John Skelhorn, Graeme Ruxton, and colleagues involved several experiments showing that predators, in this case newly hatched domestic chicks in the lab which lacked experience of real prey, misclassify caterpillars such as those of the early thorn moth (*Selenia dentaria*) as twigs rather than caterpillars.[48] In one experiment they presented one group of chicks with an unmanipulated hawthorn branch (the object being mimicked) that the caterpillars were thought to resemble, and another set of chicks with a hawthorn branch that had been bound with purple thread to change its appearance, and allowed the birds to inspect and interact with their respective objects for two minutes (Figure 29). The team then presented birds from each group with the real caterpillars and found that the chicks took longer to attack the caterpillars, and were more hesitant in

FIG. 29. Experiment testing masquerade using chicks, and caterpillars that mimic twigs. Experiments like this trained chicks with either normal branches or those bound in purple thread to change their appearance, and then tested how likely chicks were to attack real caterpillars. Chicks trained with normal branches were less likely to consider the caterpillars as prey than individuals who had experienced the branches bound with thread.

Images John Skelhorn

handling them, when they had previously seen the normal hawthorn branch, compared to those birds that had seen the manipulated branch. There was no question that the birds in each group saw the caterpillars because they were presented against a white background, so the results cannot be explained by differences in detection. Instead, birds familiar with real twigs had learned that they were of no value and subsequently responded to the caterpillars as if they were twigs.

Skelhorn and colleagues also showed that twig-mimicking caterpillars were more likely to be attacked when they were presented to chicks alongside their models (twigs) than when presented alone. This shows that when predators can directly compare the model and the mimic in the same place, they are better able to determine the presence of a prey item. This makes sense because directly comparing two similar objects side by side is easier than having to rely on memory for what one object looks like. Note that this instance will not arise with all masquerade species, however. Twig-mimicking caterpillars are found on the branches of the trees that they resemble, but this is not the case for species that resemble isolated objects such as bird droppings. It would be an unlucky bird-dropping mimic that sat in a place right next to a real bird dropping. Skelhorn and Ruxton suggest that masqueraders which are normally viewed alongside their models should, therefore, be under selection to evolve better mimicry compared to species that are normally viewed in isolation. This is an intriguing possibility, but one yet to be tested.

Surprisingly, although this work has gone a long way to establishing that masquerade does indeed work by preventing recognition and not detection, few studies have analysed how closely masquerading prey actually match the appearance of their models with regard to the vision of their predators. What's more, experiments of masquerade with more natural circumstances and predators are uncommon, although there is some evidence that web-building spiders mimicking bird droppings gain protection from foraging birds.[49] Another recent study has explored the coloration of a remarkable gliding lizard (*Draco cornutus*) from Borneo.[50] This species lives in several forest habitats, including mangrove and lowland forests. Danielle Klomp and colleagues from the Universities of New South Wales and Melbourne analysed the colour of the membranes that the lizards use to glide from tree to tree. They found that the coloration of the membranes closely matched the appearance of falling leaves from the lizards' habitat, as would likely be seen by an avian predator's visual system. Lizards from mangrove environments have colours that match local falling leaves (often bright red) more closely than leaves from lowland forests (often black to yellow-green colours) (Figure 30). The opposite is true for lizards from the lowland populations. This suggests that these two sets of lizards have diverged in appearance to resemble falling leaves in their local habitat as they glide. This could represent a considerable survival advantage, because it is well known that movement of animals is a key giveaway to predators. The lizards glide frequently, potentially putting them at great risk. As yet, no study has tested whether birds are fooled by this resemblance, or whether the gliding behaviour of the lizards has a similar motion to that of a falling leaf, but the lizards are at least of a similar size to the leaves.

There's another major type of camouflage we have not yet discussed, and it's one that is very different from the others in that it does not prevent prey being detected, but rather prevents them being captured when moving. It is a strategy called 'motion dazzle' because of the way it's thought to dazzle or confuse the observer's (specifically predator's) visual system. Generally, motion dazzle is thought to involve high-contrast markings arranged as stripes or banding patterns on an object, which make it hard for an observer to judge how fast, and in what direction, the object is moving. The roots of this theory actually stem more from military than from natural camouflage, but again it was the outcome of work conducted by a mixture of artists and biologists,

FIG. 30. Leaf mimicry by gliding lizards. The top images show the gliding flaps of several species and populations of *Draco* lizards next to falling leaves from their location. In each of the six pairs of photographs the lizard is on the left and the leaf on the right. The two right-hand lizards are from Klomp and colleagues' study.

Top images: Danielle Klomp; bottom image: Devi Stuart-Fox

including Thayer and a host of others, before, during, and after the First World War.[38,51] Those who are familiar with the way that warships and some civilian vessels were painted during the First World War may recall that the ships were often painted with elaborate geometric zigzag-type markings of high contrast, including strange shapes and stripes. One of the purported effects was to make it hard for enemy submarine targeters to judge the way in which a ship was moving. Whether or not such patterns actually worked is unclear,[38] but it is perhaps unsurprising that the theory has been applied to animals too, given how many species have stripes or zigzag markings, including fish, snakes, and of course, zebra.

So is there any evidence that motion dazzle works? Studies in the 1970s and thereafter showed that there were associations between the way that some snakes move and the patterns that they had, including longitudinal and transverse stripes and zigzags. For example, in some species, such as adders, males can look quite different to the females, often having more prominent stripe patterns, and these may be linked to higher levels of movement and activity. The implication is that colour patterns may have evolved to make prey capture more difficult. However, this is not clear-cut because it has also been shown that patterns such as zigzags on vipers seem to indicate to predators that they have a nasty bite (a warning signal). So, aside from a few indirect studies and anecdotal reports of observers being fooled by snake movements and their patterns, there was until recently no real evidence for motion dazzle working as a mechanism.

A few years ago, Graeme Ruxton and I along with various colleagues decided to test whether, in principle at least, stripy or zigzag markings could make it hard for an observer to capture a moving target.[52] To do so, we presented humans with moving rectangular targets on computer screens against images of natural backgrounds (dead leaves or grass), with the targets marked with patterns that were stripes of different widths and contrasts. We then asked humans (mostly undergraduate students) to play computer 'games' in which each person had to try and capture as many targets of a given pattern as they could in one-minute trials (Figure 31). The results showed that uniform conspicuous white targets were easy to capture, whereas capturing targets with stripes or zigzags was more difficult. In our studies we also showed that dazzle was not just an effect of having any old pattern, because targets that were hard to detect when stationary against the background (background matching targets resembling the shape and colours of the background), were much easier to capture than stripy targets when moving. Subsequent work by Nick Scott-Samuel and

FIG. 31. Computer 'game' experiments, in which human participants are asked to try and capture moving targets across a screen marked with different patterns, have been used to test the idea of motion dazzle.

Image Martin Stevens

colleagues at the University of Bristol also showed that some dazzle patterns seem to cause people to make mistakes in estimating which of two objects with different patterns is moving faster.[53] So, although the results of these and other studies are not always completely clear-cut, there is mounting evidence that dazzle markings can fool human observers at least into judging movement, with knock-on effects for capture behaviour.

The key question of course is whether motion dazzle works outside a human context and in the wild. Sadly there has been little experimental testing of this so far, and it is not an easy thing to do. Nonetheless, a recent study shows that it could work. Martin How from

Bristol University and Johannes Zanker from Royal Holloway University in the UK made a computer model of how motion detection might work in animal vision, including in a potential predator.[54] They then presented the model with images of zebra and unpatterned horses, and simulated them as if they were moving. The model suffered serious levels of erroneous information about motion when presented with the zebra, especially with images of zebra in herds. In contrast, little such motion confusion was detected when the model was presented with horse images. Undoubtedly, there are many stripy animals in nature that show high levels of movement, often species that live in groups like shoals of fish, and I am confident that motion dazzle is a key reason for this in at least some of them. But it is still an open question as to how important it is in the natural world. Note that in the case of the zebra, considerable debate still exists as to why they have stripes. Recent work by Tim Caro at the University of California, Davis and colleagues tested the ecological factors that coincide with the presence of stripes in different species and subspecies of horse, including predation pressure, temperature, and habitat type.[55] They found little evidence of stripes having evolved under selection to prevent predation from animals like lions or hyenas, but clear links between climate variables (temperature and humidity) that consistently predict the distribution of nasty biting flies, which zebra seem especially vulnerable to. It therefore seems that zebra stripes function to prevent biting fly attack, and it is possible that this is induced partly by dazzle-type effects against these flies rather than predators, although that requires further investigation.

Overall, one of the key messages from this chapter is that camouflage is not a simple one-way route to preventing detection and attack by predators, but rather a rich, intricate, and varied set of approaches used by animals to deceive predators and their vision. Considerable progress has been made recently in understanding this: in the space of just over a decade camouflage has gone from being a somewhat dormant area of research, with many theories dating back to the late nineteenth century going untested, to a subject that has made substantial advances in understanding how deception and predator–prey relationships work, and how diverse they are. Many of these theories come from some of the pioneers of evolutionary biology and early natural historians. Ironically, as Roy Behrens from the University of Northern Iowa (who has spent many years studying art and camouflage, including the works of Thayer) has noted, when Thayer was alive he was better known for his paintings of subjects such as angels, but mostly ignored by the scientific community for his ideas on camouflage, whereas today his camouflage theories are widely valued and instrumental in guiding modern-day research on concealment yet his art is largely forgotten.[56] In the past few years there have been several exhibitions of Thayer's work, focussing especially on his paintings of natural history and camouflage. I was lucky enough to give a talk at one of these, a symposium on camouflage and the life and work of Thayer at the Army and Navy Club in Washington, DC, an event partly organized with the Smithsonian Institution, discussing the achievements and resurgence of his ideas. His theories, alongside those of Poulton, Wallace, and Cott, have been proven over a century later to often be remarkably accurate and astute. We now need to devote more time and effort to understanding how things like disruptive coloration, masquerade, and motion dazzle work and evolve in real species. Many discoveries have been made in artificial situations—with computer games, laboratory studies, or experiments with artificial prey—but as yet we know comparatively little about how camouflage functions, develops, and evolves in real animals in different habitats. We also have much more to learn about how camouflage functions in a non-visual sense,[57] and in different groups of organisms, such as plants.[58] It is a subject with much more yet to tell us about the mechanics of evolution.

5

A SPIDER IN ANT'S CLOTHING

· · · · · · · · · · · · ·

Not all animals are good to eat. Instead, a great variety possess dangerous or unpleasant attributes to ward off potential attackers. The infamous venom of the blue-ringed octopus, the nasty taste of ladybird beetles, or the rancid smell of skunks are just three examples from thousands. Such defences are aimed at the natural enemies of these animals, especially potential predators, but rarely are they hidden away and concealed. In fact, quite the opposite; toxic or dangerous animals frequently advertise that they should be avoided by displaying striking patches of coloration or having unpleasant smells. Many ladybirds are bright red and yellow, skunks have conspicuous black and white stripes, and the blue-ringed octopus is marked with, well, blue rings. The reasons underlying this have been known for many years. When Darwin was formulating his theory of sexual selection, realizing that many animals were brightly coloured to attract partners, he was troubled by something his theory could not explain: why was it that many sexually immature animals, such as various caterpillars, were gaudy and conspicuous even though they did not reproduce? Darwin sought the advice of Henry Walter Bates, another key Victorian naturalist-explorer whom we will return to shortly. Bates directed Darwin to Wallace, who resolved this problem in a letter to Darwin in 1867, suggesting that brightly coloured larvae display signals telling predators that they are unpleasant to eat, much to Darwin's pleasure: 'Bates (H. W.) was quite right; you are the man to apply to in a difficulty. I never heard anything more ingenious than your suggestion.'[1] Wallace's insight was to suggest that should toxic or dangerous animals look like harmless ones, there would be nothing to warn a would-be attacker that they should be avoided, and they might be killed before the predator learns about the defence.[2]

Alongside Wallace, it was E.B. Poulton who, among the early evolutionists, set out in most detail how warning signals, or *aposematic* signals as he defined them and as they are often called today, work.[3] Poulton noted that many unpleasant species in nature possess bright colours and patterns, especially reds and yellows, and argued that defended animals would avoid investigation and attack by predators if they use these signals to advertise their unpleasantness. Warning colours, he suggested, being distinctive and conspicuous, should work by 'educating' potential enemies, helping them to learn and remember which animals should be avoided. A wealth of research spanning many decades has supported Wallace's and Poulton's ideas, showing that aposematic signals broadly work in two main ways. First, by conspicuously standing out against the background environment (and other undefended animals) toxic prey cause predators to be cautious about attacking them. Second, when predators do attack, and end up with a nasty surprise, they quickly learn to associate bright colours, patterns, smells, and sounds with toxicity or danger, and so avoid future interactions with the prey and similar species. In the process, both predator and prey benefit; the prey animal avoids being attacked or eaten, and the predator avoids a confrontation that might be costly too.[4]

Warning signals are not deceptive. In fact, they are quite the opposite, being, overall, an honest way in which unpleasant-tasting prey animals can tell predators to avoid them. But as we know by now, wherever there is an honest signal, there's a cheat as well, and such instances can tell us much about evolutionary processes and communication among species. In the case of warning signals, many completely harmless animals display colours and patterns that resemble species of toxic prey, in what is called Batesian mimicry, named after the aforementioned H.W. Bates (Figure 32). Bates was one of the great naturalists of the Victorian era. After the pair met in Leicester, he was Wallace's travelling companion when they explored the Amazon

FIG. 32. Examples of Batesian mimicry by harmless insects. Top left shows a wasp with a nasty sting, mimicked by a range of harmless species, including a hoverfly (*Fazia micrura*, top right), the raspberry crown borer moth (*Pennisetia marginata*, bottom left), and the sugar maple borer beetle (*Glycobius speciosus*, bottom right).

Top-left image © Jo Seabra 123RF; top-right image Andrew Young; bottom-left and bottom-right images Michael Runtz

and Rio Negro in 1848 to discover new species and address the question of the origin of species. For some reason that is not entirely clear, Bates and Wallace went their separate ways in 1850, to explore different parts of the region. Bates was an outstanding entomologist, and stayed in the Amazon until 1859, sending back over 8,000 species new to science. During his travels he made a number of studies and observations of butterflies, including on *Heliconius*, now famous as a group of butterflies whereby different sets of toxic species resemble one another. Bates noted that these groups of butterflies were often closely matched in both appearance and flight behaviour by other unrelated species of quite edible butterflies, but that both groups seemed to be avoided by birds. In the field, on first inspection Bates was himself frequently deceived by these mimics. After its publication, Bates was bowled over by Darwin's *On the Origin of Species*, and soon realized that the resemblances he was observing could provide some of the best evidence for evolution and natural selection. He appreciated that if a perfectly edible prey species could evolve and look like a co-occurring dangerous species then it would gain protection because predators would mistakenly avoid it too, especially when the imitation was particularly close.[5] This is the essence of Batesian mimicry, whereby a harmless species resembles a dangerous or unpleasant one. Bates' theory did indeed provide some of the best early evidence for Darwin's theory of natural selection, something that Darwin acknowledged after reading Bates' 1862 paper. In November of that same year he wrote to Bates: 'In my opinion it is one of the most remarkable & admirable papers I ever read in my life.'[6] Despite its wide acceptance by biologists today, there remain various unresolved questions regarding exactly how Batesian mimicry works, which scientific research is only now starting to solve.

One of the fundamental difficulties in studying Batesian mimicry is in showing that an animal genuinely is a mimic. For decades the subject was dominated by examples based on human subjective judgement only, with species deemed to be mimics because they looked similar to something else. But this has posed numerous problems. First, human vision is not really what matters and many animals, including the predators of most mimics, have visual systems and perceptions that are very different from our own. Second, it is often unclear what the species being copied (the 'model') actually is in many cases of Batesian mimicry. For example, many hoverflies look like bees and wasps, but which species exactly? In fact, many mimics might not even evolve to resemble one specific species, but rather to have a broad resemblance to many toxic models of a particular group, for example being a mimic of bumblebees in general. These issues make it difficult to test mimicry because if we do not know what the model species is then we cannot compare how similar model and mimic are, and so design experiments to test how mimicry works in real species. Ultimately, tests of mimicry need to show that the mimic is misclassified as the 'wrong' species by the predator (e.g. a bird mistakenly categorizes a hoverfly as a wasp), and this has not often been done. Nonetheless, progress has been made in several animal groups, and these have also taught us some important things about how adaptation and species interactions work.

The challenges with demonstrating Batesian mimicry and how these issues can be resolved are exemplified by studies of hoverflies. Hover-flies are one of the most common garden insects in temperate regions such as the UK and large parts of North America, and are found everywhere except Antarctica, with over 6,000 species described worldwide. Many are well known for being masters of flight, capable of rapid darting movements and hovering in mid-air with their rapidly beating wings. A variety of species show some degree of affinity with the appearance of wasps and bees, despite being completely harmless (Figure 33). Owing to their abundance and the large number of species, hoverflies have long been a valuable group of animals for the study and

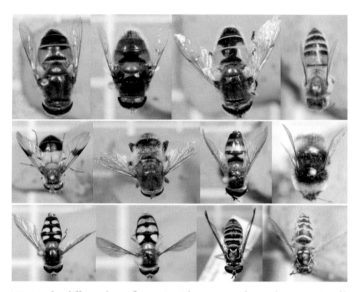

FIG. 33. Mimicry by different hoverfly species. The top row shows three species of hoverflies and their suggested type of model (a honeybee, right). The middle row shows three hoverflies thought to mimic bumblebees (right). The bottom row shows three hoverfly species thought to mimic wasps (right). Note that the model species shown here are not necessarily the specific model for each hoverfly species shown.

Images Heather Penney and from the Canadian National Collection of Insects,
Arachnids and Nematodes

discussion of Batesian mimicry. However, most early work simply involved describing how hoverflies and their putative models looked to humans, and experimental evidence that they really are Batesian mimics was sparse. One significant exception was the work of Gerhard Mostler, undertaken back in 1935.[7] He released hoverflies and their models (bees or wasps) into a room and allowed them to be attacked by free-flying birds of varying experience of different species. By changing the order that he released either hoverflies or bee/wasps, he could see how the birds responded to each type based on recent experience. Mostler made a number of interesting discoveries. In the first instance, he showed that hoverflies were protected from attack due to their mimicry, and that the degree of protection was linked to their similarity to the models (in terms of human vision at least). His work also illustrated that birds do misclassify hoverflies as wasps. When wasps were presented first, and hoverflies afterwards, the hoverflies were sometimes avoided. In contrast, when hoverflies were presented first, and wasps second, hoverflies were frequently attacked. In fact, in this scenario the birds often subsequently attacked the wasps, showing that Batesian mimicry can carry a cost to the model. Partly for these reasons, we often assume that mimics should be less common than the models for deception to be maintained, something that Bates and Wallace again suggested very early on. While many of the mimetic butterflies Bates observed did seem to be rare, it is not clear how widely this is true, and hoverflies are certainly very common at times.

Aside from Mostler's work and the occasional other exception, it was not until after the turn of the millennium that scientists really got to grips with demonstrating that hoverflies mimic various characteristics of wasps and bees. This work also began testing core theory regarding what we expect to happen over time and space regarding the dynamics of mimicry. Some of this initial work still relied on humans, but began to further demonstrate aspects of how mimicry works.[8] One such study presented university students and school children with images of venomous stinging wasps and bees, and harmless hoverflies, in addition to other species of fly not thought to be mimics. The students were, as expected, often fooled by the hoverflies, considering them as being likely to sting much more often than the non-mimetic flies. However, they were less likely to consider the hoverflies as potentially harmful than the genuine bees and wasps, showing that the mimicry is not perfect. The students were also more likely to be fooled by the hoverflies when their mimicry had already been judged by the authors of the work to be better. Thus, to humans at least, hoverflies are often misclassified as their potential models, but how good the deception is varies among species. In addition, it was not simply the presence of yellow and black markings that mattered in the judgements, but also how the markings are arranged, showing that pattern matters too in being a convincing mimic.

This study is useful because humans do after all have a good sense of vision, and so should be a reasonable judge of the closeness of any resemblance. However, work like this has rightly been criticized because humans are not the natural predators of hoverflies and wasps/bees, and therefore their visual abilities are not the selective pressure that has led to the mimicry evolving. In particular, human vision deviates in a number of important aspects from that of birds, which are likely to be the main predators involved, and this is something we'll come back to shortly. At least one earlier study did analyse how birds might perceive hoverflies. This involved a clever experimental approach by Winand Dittrich from the University of Exeter and colleagues from Nottingham University back in 1993.[9] They presented 'retired' racing pigeons with photographs of wasps, hoverflies, and other flies using a classic type of experiment in psychology called operant conditioning. The technique involves training a subject to discriminate between stimuli with rewards (or punishments) based on what decision they make, and then testing their response later to new stimuli that they have not encountered before to determine how the subject classifies them. The technique was largely introduced in the 1950s, with famous experiments training rats to press buttons associated with different-coloured lights, with one colour rewarding them with food but another choice punishing them with a mild electric shock. Unsurprisingly, the rats quickly learned to associate the behavioural choice (pressing one button type) with a subsequent outcome. In the case of Dittrich's study, the researchers trained pigeons with food rewards to recognize either wasps or flies not thought to be mimics from photographic slides (what they were trained to recognize depended on the individual pigeon). After the training phase, they presented pigeons with new images of wasps, non-mimetic flies, and photographs of hoverflies, and measured the proportion of times that the pigeons responded to the slide as a wasp as a measure of how effective the hoverfly mimicry was. They showed that pigeons were much more likely to misclassify hoverflies as wasps than they were to consider non-mimetic flies as wasps. Thus, mimicry seems to fool birds too. In addition, the categorization made by the pigeons was broadly similar to human assessment of the mimicry, although there were some exceptions that we'll return to soon.

The outcome of Dittrich's pigeon experiment might seem rather obvious, but it is important in demonstrating that hoverflies really do mimic wasps to a relevant observer (a bird), and in understanding how closely hoverflies do this. However, we still need to be cautious about the findings because the slides used in that work were photographs that represent colours in terms of human vision. Photographs are designed to replicate the colours and brightness of objects in the world by stimulating our visual system in an analogous way to that which would occur when we would naturally view such objects. As we discussed in Chapter 2, humans have colour vision involving just three cone types (often referred to as 'red', 'green', and 'blue' for the colour sensations they give rise to), whereas birds have four cone types that seem to be involved in colour perception, including one type of cone sensitive to UV light.[10] Therefore, many birds likely see a broader range of colours than we do, and may have different perceptions of the hoverflies and their models. Photographs and many TV monitors base their colour reproduction on a limited number of light-producing phosphors (red, green, and blue) or pigments. The intensity of these combines not to mimic the precise natural light spectrum of a real object in the environment, but to stimulate the cells in human eyes in much the same way as that object would when being viewed directly. For birds, however, an avian TV would need four such phosphors, including one in UV to replicate natural colours appropriately. When birds look at photographs or TV screens, we might imagine that it would be like us watching our favourite programme but with the red phosphor tuned off: that is, it would look unnatural.

Dittrich's study also did not test which specific features (e.g. colours, stripes, and so on) of the flies and wasps pigeons used to decide whether they were model, mimic, or neither. That is, what aspects of hoverfly appearance make them effective mimics. Identifying these key features is no simple task because one cannot simply ask the pigeons to tell us why they grouped the insects in the way they did. The obvious experiment would be to modify specific features of the hoverflies' appearance to make them more or less wasp like, and then see how this affects the categorization by the birds. As yet, nobody has done this. A different approach, however, has been used recently by Tom Sherratt at Carleton University in Canada and his team (including one of the original researchers on the pigeon work).[11] They took some of the original slides used by Dittrich alongside the data on the pigeons' responses. They then built a mathematical model to analyse the different features of the insects and how well each one explained the pigeons' behaviour. The characters they measured corresponded to seventeen aspects of the insects' appearance, including things like antennal length, aspects of body shape and width, and the number and colour of stripes, among many others. The model was essentially a type of 'artificial neural network'—mathematical tools with functions that use machine learning to categorize and classify stimuli and data with increased experience or training, and are broadly inspired by real animal nervous systems because they are made of a series of interconnected 'neurons'. The authors fed the appearance data into the model so it could 'learn' about the insects' appearances, and then, as they put it, 'reverse engineered' the discrimination process that the pigeons seem to have used to calculate how well different visual features determined their choices.

The results were highly revealing. First, the model showed a very close similarity to the pigeons in classifying hoverflies as either wasps or flies. Second, the most important features used were the number of stripes, the length of the antennae, the contrast of the stripes, and the colour of the stripes/body. Thus, only a subset of potential features seem to be used in the categorization process. However, which ones were most important depended on how the pigeons in Dittrich's work were originally trained. Pigeons initially trained to recognize wasps seemed to have learned to categorize them primarily based on stripe colour, contrast, and number of stripes, whereas pigeons originally trained to recognize non-mimetic flies used antennal length and aspects of body shape (in addition to colour) most. Thus, pigeons trained to recognize wasps learned most about colour and stripes, whereas pigeons trained to recognize flies used aspects of body morphology. This shows that predators learn to recognize the salient features of stimuli that are most relevant in categorizing them as a given object type, and this has knock-on effects for mimicry because mimics should be under most selection pressure to imitate those aspects of appearance that promote miscategorization as the wrong object type. For example, if you want to look like a wasp, you need to first match the colour, contrast, and number of stripes.

One of the other striking features about hoverflies is the noise that they make. If you listen to them when they are on or inside plants they frequently emit loud bursts of buzzing noises. These have been widely suggested to mimic the characteristic sounds made by wasps and bees. At present, however, the evidence does not really support this idea. Another study from Tom Sherratt's lab compared sounds made by wasps, bees, hoverflies, and non-mimetic flies for similarities in profiles.[12] Although there were some similarities between the noises produced by bumblebees and hoverflies, in general the sounds produced by wasps and bees were quite distinguishable from those made from hoverflies, and there was no evidence that the sounds made by hoverflies were more similar to their suggested models than to other wasp/bee species. In fact, all the hoverfly species tended to sound quite similar to one another. This is perhaps not where the idea ends though, because the sounds investigated were limited to those made when the insects were under simulated attack. It is possible that sound mimicry occurs with regard to noises the insects make when not being attacked, or that tethering the insects in chambers to measure their sounds altered them in some way. Indeed, ideally mimicry and warning signals should prevent attacks from occurring in the first place, rather than acting as secondary defences once an attack occurs, and so we might expect mimicry to involve sounds used prior to attack. Finally, as the authors note, we don't know how the sounds would be analysed by the hearing of the predator, and so it is difficult to be sure what aspects of sound mimicry might matter.

While acoustic mimicry by hoverflies is controversial, it does seem that many species complement their colour pattern mimicry with behavioural mimicry. Here, certain hoverflies apparently adopt behaviours that increase their resemblance to a wasp or bee. This includes holding their front legs above the head in a manner resembling the elongated antennae of many wasps, and simulating the movement of an abdomen when a wasp stings. Recent work, again by Sherratt and others, has tested whether behavioural mimicry goes hand in hand with colour pattern mimicry.[13] They collected live specimens of fifty-seven species of hoverfly from Canada and tested their behaviour in the lab. To do this, they prodded the flies with the beak of the stuffed head of a bird (a blue jay) to simulate an attack, and then recorded the different behavioural responses. In addition, for each hoverfly species they compared the level of visual mimicry of the hoverfly to one of five different wasp and bee models, here based on human assessment of how good the mimicry was. Of all the species analysed, just six showed some degree of behavioural mimicry, with all of them being hoverflies that likely mimic wasps. At the moment though, how often such behavioural mimicry has evolved independently in different groups remains to be seen.

Studies of hoverfly morphology have gone a long way to demonstrate that mimicry really is the adaptive explanation for much of their appearance and behaviour in many species. Yet there also exist several other predictions about the existence and relative frequency of models and mimics both over time and space that have been tested on hoverflies. One of the simplest is that frequencies of mimics in the wild occurring in different geographic areas should correspond to the local abundance of their respective models. Testing even such a simple idea has proven difficult though, not least because, as we discussed earlier, it is notoriously difficult to be sure which hoverfly species is mimicking which model. However, recent work has started to address this question. Malcolm Edmunds from the University of Central Lancashire has been studying mimicry and defensive coloration in animals for decades, and his name is frequently linked with many of the main ideas behind Batesian mimicry. Along with Tom Reader from the University of Nottingham he conducted an eleven-year study of hoverfly and model frequencies in fifty-two locations across the UK.[14] They took advantage of the fact that one UK species of hoverfly that mimics bumblebees, *Volucella bombylans*, is also polymorphic; that is, individuals come in one of a number of different colour forms (just like some camouflaged species). These forms appear to mimic different species of bumblebee, with some of the hoverflies resembling black and yellow bees and others mimicking bees that are black with a red 'tail'. The scientists measured the frequency of both hoverfly morphs and the different bee species, with the simple prediction that the respective hoverfly morphs should be more common in locations where their own bee models are relatively abundant. This is largely what they found, with at least some of the different hoverfly morphs showing frequencies that correlated with those of their bumblebee models. There is one factor that might complicate this example, in that *Volucella bombylans* is also thought to be an invader of bumblebee nests (a type of parasite we'll return to in Chapter 7), and so its mimicry might also have evolved to evade the defences of nesting bees. Indeed, this was a source of some debate between Poulton and others back in 1892. Despite this caveat, other work has also shown that the activity patterns of hoverflies, in terms of what time of day they are most active, also often coincides with that of their putative model.[15] So while it's still early days regarding this sort of evidence, it is consistent with predictions for Batesian mimicry.

Studies of hoverflies have been valuable in starting to disentangle the dynamics and mechanisms of how Batesian mimicry works. However, one of the major missing pieces of evidence remaining is in showing how predators respond to hoverflies in more natural situations. In particular, we need evidence showing that differences in the level of mimicry really do make predators more or less likely to attack. Thankfully though, another mimicry system has done this very effectively.

As we know by now, Wallace was full of ideas about deception and mimicry. He was one of the first people to suggest that various jumping spiders mimic ants.[16] It's not hard to see why, because the resemblance can be truly striking. Some spiders' entire morphologies are modified in shape and colour to match that of an ant, with fewer hairs, narrower body segments, and an elongated body, with one pair of legs often held forwards in order to resemble the appearance of antennae. What's more, when threatened, ant-mimicking spiders will often adopt behavioural mimicry, including ant-like threat displays. So remarkable to

FIG. 34. Ant-mimicking jumping spiders. Some of the most common types of ant mimics are jumping spiders, and many invertebrates have evolved to look like ants, often in order to avoid predation because ants are well defended and shunned by predators.

Images ownza/shutterstock.com and © defun/Istock.com

human eyes can the resemblance be that the only clear giveaway is the eight legs when you count them (Figure 34). So why might spiders benefit from mimicking ants? There are likely to be several reasons. One is that it could give them access to nests and enable spiders to sneak up and eat the ants: another type of aggressive mimicry. But more often it seems to be for protection from predators, including both birds and other spiders. Ants are notoriously unfavourable prey. Not only do they have strong defences like bites and stings, but they can also produce nasty substances like formic acid, meaning that they often don't taste very nice. They also exist in large numbers, allowing for group defences. Mimicry of ants may have evolved over seventy times in arthropods, most often in spiders but also in bugs, beetles, and things like parasitic wasps. In all, around 2,000 species of ant mimics have been described, and they are especially common in the family of spiders called Salticidae, or jumping spiders. In fact, one genus of jumping spider (*Myrmarachne*), with 200 known species, comprises exclusively ant mimics.[17]

Ximena Nelson and Robert Jackson (mentioned in Chapter 3 in connection with *Portia* aggressive mimicry) from Canterbury University in New Zealand have conducted extensive experiments to test whether spiders really do mimic ants, and how this works. One of the main predators of ant-mimicking spiders is in fact other spiders, and doing controlled experiments with these in a lab is therefore easier than with, say, birds. It has taken a very long time since Wallace's suggestion for controlled experiments to show that ant mimicry is genuine, but work by Nelson and Jackson in the mid-2000s onwards has provided compelling evidence. In one experiment, they presented predatory jumping spiders (which have excellent vision to detect and analyse prey and potential threats) with a range of mounted dead arthropod specimens, and gave them a choice of whether to approach them or move to a different part of the experimental arena.[18] Spiders showed no evidence of avoidance of non-ant-like prey, yet often moved away from ants or ant-mimicking spiders.

The above experiment did not just demonstrate that ant mimicry exists and works, but also something else: that avoidance of Batesian mimics can be innate rather than learned. In many studies of ant mimicry, including here, the predator had not encountered ants or ant mimics before and was naive to them. Their avoidance cannot, therefore, be based on learning. So, although we often think of Batesian mimicry as something that involves a predator using prior experiences to learn subsequently to avoid both dangerous prey and their mimics, it can also be 'hardwired' over the course of evolution. When the danger is strong enough and the stimulus is common enough, as is the case with ants being so numerous, then it may be beneficial to possess an unlearned avoidance of these and other similar-looking animals.

Ant mimicry seems effective, but is there any evidence that the level of mimicry affects how strongly predators avoid the mimics? That is, are more closely matched mimics avoided more? In another study, Nelson offered ant-mimicking spiders to a predator, in this case *Portia fimbriata*.[17] She used individuals from several spider species with different levels of ant mimicry (judged by human eye), from either very good to seemingly inaccurate, and recorded how *Portia* responded. As expected given their excellent vision, *Portia* showed variation in avoidance behaviour in line with the level of mimicry, with avoidance being stronger when the mimicry was judged to be more similar by humans. As such, the level of signal accuracy is important in how effective the mimicry is in preventing attack. It makes it even stranger that so much variation in the accuracy of mimicry exists, because there should be an advantage to evolving better mimicry. We will return to this issue of 'imperfect mimicry' shortly.

One of the odd things about some ant-mimicking spiders is that males and females sometimes look different, with mimicry by males being generally worse in such cases. This can be due to males owning an enlarged pair of front appendages (chelicerae) (Figure 35). Remarkably, Nelson and Jackson found evidence that male spiders do not seem simply to mimic ants, but specifically copy encumbered ants carrying objects (such as prey items or nest material) in their mandibles.[19] Quite why males have opted for another appearance is unclear, but the enlarged chelicerae may have arisen from selection by male–male competition and threat displays or perhaps females choosing males with enlarged chelicerae. In such cases, a trade-off between successful mimicry and other functions can arise because males may suffer less effective resemblances to ants overall. In Nelson's study, where she offered ant mimics with different degrees of mimicry the predator showed less aversion to these male encumbered ant mimics than to females. In addition, other experiments by Nelson and Jackson have shown that *Portia* spiders avoid attacking both female ant mimics and encumbered males as if they were considered ants. However, another jumping spider that actually specializes in hunting ants (*Chalcotropis gulosus*) is more likely to attack real ants carrying something than unencumbered individuals, and therefore also more likely to attack male ant-mimicking spiders than females. Unencumbered ants may be

FIG. 35. Difference in appearance between male and female ant-mimicking jumping spiders. The males, such as the spider shown here, have greatly enlarged front appendages (chelicerae), which may play a role in mate choice. Work has shown that they also mimic the appearance of an ant carrying something, such as a prey item, in its mandibles.

Image mcmac/shutterstock.com

more risky to attack because they have their mandibles free to defend themselves, whereas ants carrying something cannot use them.

The level of mimicry of ants by some spiders is truly impressive, but some species take this a step further. The jumping spider *Myrmarachne melanotarsa* not only has the physical appearance and behaviour of ants (that they often associate with), but they also form groups, resembling a small colony of ants, with tens or even hundreds of individuals. Nelson and Jackson's work shows that predatory spiders are less likely to attack both ants and ant mimics than they are other spiders.[20] More importantly though, predators show greater aversion to groups of ant-mimicking spiders (or a group of ants) than to lone spiders, and they show little avoidance of groups of non-ant-like spiders. Such grouping of mimics, called 'collective mimicry', strengthens the overall defence presumably because predators are even more reluctant to attack a group of ants that could engage in collective defence. Note that this is somewhat different from the example of the blister beetle larvae we encountered in Chapter 2, in which the larvae work together collectively to form the characteristics of a single object (a bee), rather than mimicking individuals of a group as these spiders do.

Just like camouflage, most work on Batesian mimicry has focussed on visual signals, partly because they are often most obvious to us and often most tractable to study. Yet Batesian mimicry also occurs in other modalities, including sounds and smells. Burrowing owls (*Athene cunicularia*) nest in holes in the ground made by rodents such as ground squirrels, but which often also house rattlesnakes. In fact, all three animals can sometimes live together in the same network of burrows, and for a time it was mistakenly believed that they happily live together communally. Instead, the reality is that the snakes pose a threat to the ground squirrels, but the owls may benefit from the rattlesnakes' presence. When disturbed, the owls (hidden from view in the burrows) emit a hissing noise that sounds like a rattlesnake warning, potentially deterring some of the owl's own predators such as badgers, weasels, and coyotes. The snake and owl sounds have a similar acoustic structure, and work in the 1980s by Matthew Rowe and colleagues from the University of California, Davis compared how ground squirrels respond to the owl hisses (and control sounds), including individual squirrels from populations that naturally occur with snakes, and those from places where rattlesnakes do not occur.[21] This showed that ground squirrels familiar with snakes avoid the owl hisses, whereas those that have not encountered snakes are less cautious of the display. Presumably the owl's predators would respond in a similar manner.

Acoustic Batesian mimicry is likely to be common in bat–moth interactions too. Bats are major nocturnal predators of many moth species, capturing them both in flight and grabbing moths that sit on vegetation (a behaviour known as gleaning). Many bats use their highly sophisticated echolocation calls and hearing to detect the moths they hunt. The very high 'ultrasonic' frequencies of sound involved are often well beyond our hearing range (hence the need for an echolocation detector to hear bats hunt and navigate), and can be so refined in some bats that individuals not only use them to navigate and avoid objects, but also to detect tiny and refined features of their prey. For example, some bats are capable of detecting the texture and even the wingbeat frequency of a flying insect, allowing them to discriminate very precisely between species and prey types, and enabling them to learn about what they are hunting. As with many predator–prey relationships, prey have evolved defences. One of the most important is the evolution of hearing organs, which occur in various places over the insects' bodies (even on the legs), resulting from multiple independent evolutionary origins. These ears are precisely tuned to the high-frequency calls of bats, allowing the insects to take evasive action, such as changing direction or suddenly dropping to the ground so they cannot be caught.

In addition to hearing bat calls, many insects have also evolved an ability to produce ultrasonic sounds themselves, in the form of clicks and pulses that they use as defences. Again, this ability seems to have evolved several times independently. The mechanisms by which ultrasonic insect sounds are produced varies, from the use of special plate-like organs on tiger moths called tymbals, to beating together wings and wing cases in tiger beetles. In tiger moths, some species signal to bats that they are toxic, being a form of acoustic aposematism. Bats can learn to avoid these defended moths based on the type of clicking noises that they make. Some completely harmless moths also produce ultrasonic clicks when they hear a bat approach, and these mimic the clicking sounds of toxic species.

Experiments at Wake Forest University in North Carolina by Jesse Barber and William Conner provide probably the most convincing tests of acoustic Batesian mimicry so far.[22] They trained captive bats in sound-controlled rooms to attack toxic sound-producing moths, and used high-speed infrared cameras to record their interactions in the dark. As expected, bats quickly learned to associate the clicking sounds of the moths with toxicity, and avoided them after a few trials. Next, they offered the bats palatable moths that also made noises similar to the toxic species. Here, the bats still avoided the moths despite them being edible. Only after a number of encounters did some of the bats realize that the moths were edible and begin attacking them. At present, how common acoustic mimicry is in this type of scenario remains largely unknown, but as the authors note in their report there are over 11,000 species of tiger moth, not to mention many other groups of insect that could respond to bat attacks with acoustic displays. Acoustic mimicry is therefore surely much more common than presently known. In fact, one recent study found that a species of Geometrid moth (a group only distantly related to tiger moths) called the orange beggar (*Eubaphe unicolor*) has evolved a sound-producing tymbal organ convergent in both form and sounds to those produced by many tiger moths.[23] Further analysis has shown that although the orange beggar produces sounds very similar to some toxic tiger moths found in the same area, it is itself palatable to bats. As such, the orange beggar seems to be a species of moth that has evolved Batesian mimicry of tiger moths against bats.

At times I have alluded to the fact that there is much variation in how closely mimics resemble their models, and the presence of these so-called 'imperfect mimics' is a frequently discussed issue in Batesian mimicry (Figure 36). Wallace used the existence of imperfect mimics as

FIG. 36. Imperfect mimicry by hoverflies of wasps (far right). Hoverfly species from left to right have mimicry ranging from very poor to very good, to human eyes.

Images Heather Penney and from the Canadian National Collection of Insects, Arachnids and Nematodes

evidence against a Creator and in favour of natural processes leading to the origins and diversity of species, because it showed that species were not 'perfect'.[2] They represent a puzzle, however, because a closer resemblance by the mimic to its model should be more effective in deceiving a predator, so why do so many imperfect mimics exist? Multiple explanations have been proposed to explain this—there are more than ten different hypotheses, many of which may work together to a lesser or greater extent.[24] Some of the leading ideas are: that mimics may only appear imperfect to us because of large differences between human and predator vision; that mimics may resemble multiple models simultaneously but none perfectly (a compromise in appearance or jack of all trades); that mimicry need not be perfect if the model is highly toxic and therefore strongly avoided; that predators may only attend to some prey features and not others when assessing mimicry; and that mimics may be part of a continuing evolutionary process, with models evolving away from them in appearance, preventing mimicry from being perfect. The majority of these ideas have been tested only with mathematical models, if at all. However, recent developments mean that we are starting to gain a better idea of the key issues and which explanations seem to be most likely. These issues of imperfect mimicry have mostly been discussed with regard to Batesian mimicry, but in fact apply to any form of mimicry in which the resemblance seems less than very good.

Let's start with the theory that apparently poor mimics to us may actually be good mimics in the eyes of the real predators. This certainly seems like a plausible idea because vision does differ greatly among species, but there's actually little evidence for this explanation. The hypothesis really came to light after Dittrich's hoverfly and pigeon study. Aside from the general findings that pigeons were often fooled by mimicry, the researchers found some odd results. The most significant of these was that the pigeons classified two common species of hoverfly as being very similar to wasps, even though to humans they are poor mimics. Dittrich argued that this might have occurred if the hoverfly markings exploited some key aspect of learning or recognition in the pigeons, but discussed this idea only briefly. Shortly after, two biologists, Innes Cuthill from the University of Bristol and Andy Bennett then at Oxford University, published a note raising a question mark over some of the conclusions Dittrich and colleagues made due to limitations in the way that the photographs were presented to pigeons.[25] Specifically, as we have already discussed, the problem is that the photographic slides comprised colours designed to stimulate human vision rather than avian vision, and so the hoverflies might have looked unnatural to the pigeons. Thus, the pigeons may have classified these hoverflies very differently had all the relevant information to their vision been available.

The idea that imperfect mimicry is driven by differences between human and avian vision is an intriguing one. However, while the principle of this is undoubtedly reasonable, it is unlikely to be the whole story. Although birds do have a different visual system to us, their vision is largely thought to be excellent and, as we have noted, probably allows them to perceive a greater and not lesser range of colours than we can. Many birds also seem to have comparable abilities to humans in discriminating shapes and patterns. Overall then, if a mimic looks ineffective to human eyes, it likely looks even less convincing to a bird. Indeed, several of the scientists involved with those original pigeon studies did later undertake a small experiment presenting pigeons with dead pinned specimens and found similar results to the experiments involving slides.[26] Pigeons learned to discriminate between wasps and flies, and when presented with four species of hoverfly they showed some evidence of classifying hoverflies as wasps, with the strength of this response depending on the hoverfly species in a similar way to when photographic slides were used. Having said that, few other studies have yet analysed how birds directly presented with real hoverflies and wasps (as opposed to slides, for example) rank them, or conducted experiments showing that the birds are fooled by the real-life deception, and so until more experiments are done we cannot rule out the idea that imperfect mimicry reflects differences between human and predator perception.

Dittrich and colleagues' suggestion—that the hoverfly patterns exploited some important aspect of learning or pattern recognition by the pigeons—is an intriguing one. Several related ideas for imperfect mimicry come into play here, generally revolving around the idea that not all aspects of the mimic's appearance may be analysed by the predator, and so only some features may be under selection to evolve effective mimicry. This means that while the overall appearance of a mimic may not seem convincing to us, the key features used by predators in their assessments may be similar, and this may be enough for the deception to work. The hoverfly experiments described earlier, identifying features that predators use to classify models and mimics and discriminate between them, is relevant here. That work showed only a subset of prey features may be used in the learning and classification process (such as hoverfly stripe number, colour, and antennal length), whereas other features are relatively ignored. This is consistent with only a subset of the animal's traits mattering to the predator and hence being selected for in better mimicry. Features not used in learning and discrimination would be under no selection to better resemble analogous traits in the model, and hence while the overall appearance to us might not be especially compelling the mimicry still works well.

These arguments are consistent with another recent experiment by Baharan Kazemi and colleagues at the University of Stockholm, which took a different experimental approach by using artificial paper prey (not intended to mimic any real species) and birds (blue tits) as predators in lab experiments.[27] They trained birds to search for model prey, whereby some were 'edible' mimics that had a mealworm larva hidden underneath (which many birds love to eat), whereas others were unprofitable models that had no reward. The birds had to learn to discriminate between the rewarding and unrewarding prey based on several different aspects of visual appearance, including colour, pattern, and shape. After this learning phase, the team presented the now experienced birds with perfect mimics that matched the colour, shape, and pattern of the models, as well as imperfect mimics that matched the model but only in one attribute: just colour, shape, or pattern. This showed that some aspects of prey appearance, such as colour, seem to be especially salient in learning (they were learned faster), whereas aspects of pattern and shape are less important. In addition, predators were fooled by prey that mimicked only the colour of the models and ignored them, but they did not ignore prey that mimicked only pattern or shape. Thus, colour was more important in driving mimicry than the other two attributes. Although the artificial prey used and the aspects of colour and pattern were very simple, the results suggest that predators might only attend to certain prey features, ignoring others that then remain less important in underlying mimicry. The key thing here is not so much that colour itself was used, but rather the principle that some prey features seem to be more salient to a given predator and therefore more important in learning and categorization, and hence be the focus of selection for mimicry.

So, is there any evidence that this might be the case under more natural circumstances? David Kikuchi and David Pfennig from the University of North Carolina tested the idea that predators may only

FIG. 37. Mimicry by kingsnakes. Top images show the highly venomous western coral snake (left) and its non-venomous mimic the sonoran mountain kingsnake, both from Arizona. Bottom images show the highly venomous eastern coral snake (left) and its non-venomous mimic the scarlet kingsnake (right).

Top images David W. Pfennig; bottom-left image Wayne Van Devender; bottom-right image David W. Pfennig

attend to some features of prey appearance by studying mimicry by the non-venomous scarlet kingsnake (*Lampropeltis elapsoides*) of the deadly coral snake (*Micrurus fulvius*) in the south-eastern US (Figure 37).[28] Both species have red, yellow, and black rings, but the different order of colours along the body between the species gives their identity away; as the saying goes: 'red on yellow, kill a fellow; red on black, venom lack'.

The scientists conducted field experiments whereby they made artificial snakes from clay with the colour patterns adjusted in different ways. Some models were made to resemble the coral snake, in both the order of the banding colours and the proportion of the colour types. Then they made two mimetic models, one being a 'good' mimic with the same proportions of colours as the coral snake version but in a different order, and another being a 'poor' mimic that differed from the coral snake in both the order of the banding and the proportion of the

FIG. 38. Artificial snakes used to test how predators respond to different colour patterns and markings. These models are made from clay and have been given contrasting appearances. The one at the bottom of the right-hand image has been attacked by a predator.

Images David W. Pfennig

three colours (having more red and less black). The snakes were placed in the field and collected after five weeks with the level of attack by mammals and birds (judged by beak, claw, and bite marks) assessed for each model type (Figure 38). Kikuchi and Pfennig found that the proportion of red and black coloration was a key factor in predation risk, but ring colour order did not matter. This was because there was no difference in the level of attack between the model and the good mimic, whereas the poor mimic was attacked significantly more than the other two model types.

So why is there no selection for mimicry on stripe order whereas there is for the colour proportions? Because the coral snake models are so dangerous, it is likely that as mimicry gets better there becomes a reduction in the strength of selection for further improvements in the resemblance to the model; predators should generalize colour patterns more widely when the model is more toxic because the cost of a mistake is more severe. In such cases, mimicry need not be perfect. However, the authors argue that an additional factor is needed to explain their findings. They found that the kingsnakes need only mimic some attributes of the coral snake models, specifically the proportion of different colours but not the order of colours. If the only thing going on was relaxed selection on accurate mimicry because the coral snake model is very toxic then we would expect lower selection on all mimetic traits (both colour order and proportions), which wasn't the case. So why then might colour proportions but not stripe order be under selection? The answer may be due to time constraints and limitations on the predator to learn about several features of prey appearance. While humans can use stripe order to tell the difference between the two snakes, the process takes time because one has to directly look at the order of several stripes and interpret them. You cannot simply gain a quick overall judgement of pattern order, whereas this can be done for the relative colour proportions, which can be rapidly judged. Predators are often under severe time constraints. Not only do they have to make an accurate decision, but also a fast one (leading to a so-called speed versus accuracy trade-off), otherwise they either lose valuable time for foraging if they are slow (and their prey may escape), or they put themselves at risk with a poor choice (such as attacking a venomous snake). In this instance, it may be that the cognitive process needed to assess stripe order simply takes too long for predators to engage in this method.

The coral and kingsnake example is also valuable for providing another line of evidence for the existence of imperfect mimicry. The kingsnake occurs in a wider geographical area than its coral snake model, such that some kingsnakes coexist alongside the model, whereas other individuals occur where coral snakes are not found. We might expect selection for accurate mimicry to be stronger where the ranges of the two snakes coincide, because predators would have more opportunity to learn that the model is dangerous and how to discriminate it from the mimic. However, George Harper and David Pfennig found the opposite result.[29] At locations where the coral snake boundary ends, but the kingsnakes continue, there is strong selection against imperfect mimics, whereas where both ranges overlap, imperfect mimics are more common. Where the ranges overlap, predators are also more likely to avoid imperfect mimics than predators occurring on the edge of the coral snakes' range. The explanation seems to relate to the relative proportions of both model and mimic. When the coral snake is rare compared to the kingsnake there is reduced selection for predator learning of mimicry because a predator is unlikely to be confronted with a coral snake. This means that only very good mimics are likely to be avoided because the risk to a predator is low. In contrast, when the coral snake is relatively common (where the ranges overlap), then there is a higher risk of attacking a model, and because the model is very dangerous it pays predators to avoid even poor mimics. Thus, kingsnakes can get away with being imperfect.

Other potential explanations for imperfect mimicry remain plausible but are as yet largely untested. One common suggestion is that mimics resemble several model species at the same time, albeit imperfectly. In the case of spiders mimicking ants, Nelson suggests that inaccurate mimicry might be possible because the ant models are so aversive to predators that even relatively crude resemblance is enough to discourage predators. In addition, imperfect mimics may gain by broadly being able to mimic several ant species rather than specializing on one that may only occur in a single geographical area. This seems especially reasonable for ant mimicry, given that ants often have such characteristic body shapes. In hoverflies, however, recent work analysing the appearance of thirty-eight different species and ten putative models found no evidence that some imperfect species fall as intermediates in appearance between two or more models.[30] While helpful this does not really discount the idea, because we would also expect the accuracy of each hoverfly in resembling the model species to be influenced by the relative toxicity and abundance of the models. That is, we might actually predict that imperfect mimics are somewhat closer to some wasp or bee species than others. What the study did reveal, however, is that the level of mimicry seems to improve in hoverflies when species become larger. This is expected if there is relaxed selection on certain mimics. Small hoverflies are less profitable to predators than large ones (which offer a greater food reward), and so are less likely to be attacked and eaten, meaning that they may benefit less from mimicry and therefore be under reduced selection. In contrast, for large hoverflies it may be important to be good mimics to offset the costs of heightened attack risk because they are such valuable prey.

Before finishing our look at Batesian mimicry, it is worth pointing out a general observation—that most instances involve an invertebrate being the mimic, either of other invertebrates or vertebrates (we'll discuss some examples in Chapter 6). When vertebrates engage in Batesian mimicry they tend to resemble similar vertebrate species (such as snakes mimicking other snakes); cases of vertebrates mimicking invertebrates are very rare. When it does arise it tends to involve acoustic mimicry rather than visual appearance, perhaps because changes in morphology are more difficult to produce or impart a cost on other functions. However, when invertebrate mimicry by vertebrates occurs, it can be remarkable. In early 2015 a study was published reporting the remarkable incidence of a species of Amazonian bird (*Laniocera hypopyrra*), whose chicks apparently mimic toxic caterpillars (Figure 39).[31] Nestlings are covered with striking bright orange feathers, modified to look like long hairs, and completely unlike the dull grey adult birds. They look completely unlike the chicks of most other bird species too, and when disturbed even move their head and body from side to side, somewhat like a caterpillar walking. The authors of the study suggest that the nestlings resemble a local toxic caterpillar and rely on Batesian mimicry for protection from nest predators because their nesting period is unusually long, leaving

146

FIG. 39. An Amazonian bird (*Laniocera hypopyrra*) with chicks said to resemble a locally occurring toxic caterpillar (bottom right). They have bright orange feathers and when disturbed move in a manner characteristic of the caterpillar.

Caterpillar_Megalopygidae (Image Wendy Valencia);
Laniocera_hypopyrra_Nestling_Day4 (Image Duván García);
Laniocera_Nestling (3) (Image Santiago David Rivera)

them very vulnerable to predation (around 80 per cent of nests may fail). The resemblance of the chicks to local hairy orange caterpillars is remarkable, though at present this is all based on human assessment. We will have to await proper experiments to test this phenomenon.

Few other clear instances of vertebrates mimicking invertebrate species exist. In the 1970s, Raymond Huey and Eric Pianka suggested that a species of lizard (*Eremias lugubris*) from the Kalahari in southern Africa resembles the colour and the walking behaviour of a toxic beetle, which sprays acidic chemicals when attacked.[32] The adult lizards are relatively camouflaged and walk in a similar manner to other lizards. Juveniles, however, are black and white in colour and move in a jerky fashion, arching their backs and pressing their tails to the ground in a way that resembles the beetles. The lizards are also active at similar times of day to the beetles, and may gain protection from a suite of predators, including birds such as shrikes, mammals such as foxes and jackals, and snakes. In truth, the resemblance to my eyes is not hugely convincing, but the authors claim to have been fooled more than once in mistaking the lizards as beetles, and as Wallace frequently noted animals need to be viewed in the field to appreciate their appearance. Until experimental tests are conducted, we will just have to speculate on this example too.

As with many other areas of anti-predator coloration and behaviour, much of what we have learned about mimicry has arisen comparatively recently, despite the idea being an old and important one. The study of mimicry touches on many aspects of biology: from issues of animal vision, perception, and learning, through to aspects of evolutionary processes such as trade-offs in animal traits for different functions. Undoubtedly, we still have much to learn about the dynamics of Batesian mimicry, and perhaps especially why mimicry is sometimes so impressive, yet sometimes so poor. There exist many ideas to explain imperfect mimicry but we are still some way from uncovering the main reasons, and there are probably several of these operating in different species. What's also clear is that multiple selection pressures and factors appear to be involved in driving mimicry evolution, and these need not always be the same among different animal groups. In some cases, mimicry represents a balance between avoiding attack and other selection pressures (such as attracting mates), whereas in other cases it can be remarkably sophisticated and driven by highly discriminating predators. Mimicry also needs to be appreciated from a community perspective because the frequency and success of mimics can depend on the relative abundance of their model species, and this can even lead to the evolution of several morphs of the same species of mimic arising that mimic different models. Above all, Batesian mimicry has proved to be a wonderful testing ground for understanding adaptation and how species interactions can work, and that will surely continue for some time.

6

BLUFF AND SURPRISE

· · · · · · · · · · · · ·

When a predator attacks, one might think that prey should flee, as fast as possible. But many species don't do this. Instead, they behave somewhat erratically, flashing conspicuous body parts, running around haphazardly, or even bouncing and jumping up and down. Back in the 1960s, scientists suggested that these so-called 'protean' or 'deimatic' behaviours were adaptive, working to confuse and disorientate predators, buying additional time to escape (beyond simply making evasive manoeuvres).[1] In Chapter 5 we considered Batesian mimicry, whereby harmless animals avoid being attacked by predators through a resemblance to dangerous or unpleasant species. Batesian mimicry works through preventing an attack from occurring in the first place, a so-called primary defence. Yet animals also use secondary defences, which come into play once a predator has begun an assault. Some species bluff their way out of danger, not by mimicry but through surprise and unexpectedness, causing the predator to pause or abort its attack entirely. Other defences confuse or manipulate the attack of a predator in a manner that allows the prey to escape with minimal damage. It is these secondary defences that we will discuss in this chapter.

The most widely studied protean defences are startle displays. These involve a sudden change in appearance of an animal, for example through displaying bright colours, patterns, and sudden sounds. Normally, animals that use startle displays rely initially on camouflage (a primary defence), but then resort to startle if a predator finds them. For instance, a camouflaged moth sitting on a tree trunk might suddenly flash open its hindwings when a bird approaches, changing its appearance from dull brown to a flickering bright red and black. The general idea is that the predator is somehow confused or surprised by the startle display, such that it hesitates in its attack or calls it off entirely. Startle behaviours are seemingly widespread in various insect groups, including moths and grasshoppers, but have also been reported or studied in animals as diverse as cuttlefish and birds, and even include bright bursts of bioluminescent light by deep-sea organisms.

The key questions are: do startle displays work, and if so why do they work and how effective are they? Until the 1980s, most work on the subject was anecdotal and observational. But then a series of studies conducted by three researchers under the supervision of Ted Sargent at the University of Massachusetts tested what factors would make startle displays work. Sargent had been working on North American *Catocala* underwing moths, a group of species with wings well concealed against the trees on which they sit, but often bearing brightly coloured hindwings that might startle predators when suddenly revealed. One of the striking things about the *Catocala* moths is the diversity that exists in their colour patterns. Some species have red hindwings with black bars; others are blue, yellow, orange, or even just black and white. In fact, in the area of New England alone there are around fifty species of *Catocala* with over forty distinct varieties of hindwing appearances.[2] There's nothing apparently unappealing about the moths to predators such as birds—because Sargent showed that birds would happily eat a range of *Catocala* species. Such diversity

FIG. 40. The startle displays of underwing moths. The top species is the blue underwing (*Catocala fraxini*), showing its camouflaged appearance with the wings closed and the blue and black hindwings used in its startle display. The bottom species are the dark crimson underwing (*C. sponsa*, left), and the broad yellow underwing (*Noctua fimbriata*, right).

Images Alexandra Török

in colours among species is not restricted to *Catocala* moths either—there are several other groups of related moth and grasshopper species showing similar variation in hindwing colours and patterns (Figure 40). So added to our questions of what makes startle displays effective, we might also ask: why do such variants exist and what do they tell us about how startle displays work?

The first study to really deal with these questions was by Frank Vaughan.[3] He suggested that, in addition to the general sudden change in appearance that occurs with a startle display, several (related) factors could make them effective at deterring bird attacks. First, predators may be cautious about any colour that they have not seen previously on prey, something called neophobia. Here, a predator that is used to seeing red and orange moths may be strongly startled by a completely new moth with blue wings. Second, and similarly, predators might be more hesitant to attack a rarely encountered colour that was not expected; that is, they are taken by surprise. That is, they might have seen blue before, but it's so uncommon they were not expecting it. Third, predators may be more startled when encountering a familiar object in unusual circumstances (an anomaly). For example, if they normally encounter a blue moth on oak trees and red moths on ash trees, they may be taken aback if these switch round. Finally, birds may simply inherently avoid colours that tend to be associated with toxic prey, such as reds or yellows.

Vaughan conducted lab experiments with captive blue jays (*Cyanocitta cristata*), whereby the birds were trained to push aside twenty-four flaps on an experimental wooden board to reveal a coloured disc. The birds would then remove each disc to reach a food reward underneath (a beetle larva). Using this set-up he could measure how birds hesitated when faced with different-coloured stimuli, while controlling their experience of the various colours used. Birds were initially trained with discs of one colour only (e.g. red or blue), and Vaughan could then measure their hesitation when presented with new colours based on the time delay between pushing aside the flaps and pulling out the coloured disc to get the food. One of Vaughan's primary findings was that birds hesitated more when they encountered a new colour, but this was seemingly not influenced by colour type. As such, their hesitation was caused by novelty rather than avoidance of any particular colour type. He also found that birds sometimes paused longer when faced with uncommon colours that they had seen before.

Vaughan's experiments showed nicely how aspects of novelty and unexpectedness make startle displays effective. However, the set-up, involving pushing aside flaps on a wooden board to reveal coloured discs, is some way removed from how a bird would encounter a real moth. We can't really be sure that the birds treated the task in the same way as responding to natural prey. Shortly after, another of Sargent's PhD students, Debra Schlenoff, undertook similar experiments, but this time with more naturalistic stimuli.[4] Once again using captive blue jays in indoor foraging tasks, Schlenoff made artificial moths with cardboard 'forewings' covering plastic 'hindwings' that had been marked with different colours and patterns. When the birds pulled the moths from a presentation board to gain a food reward, the hindwings were suddenly revealed. This enabled Schlenoff to measure a range of behavioural responses to different stimuli, from increased hesitation during an attack to dropping the prey entirely and flying away.

Schlenoff's several experiments each tell us something interesting about how startle displays work. To begin with, birds that had been initially trained to attack prey with uniform grey hindwings were startled when they saw novel brightly coloured patterns, and this response persisted for several days. However, the reverse was not true; birds didn't show greater hesitation or avoidance of grey winged moths if they had been trained on colourful stimuli. So, while novelty is important, how conspicuous and contrasting the display is also matters. Next, Schlenoff showed that once the birds had become familiar with and habituated to one hindwing colour pattern, they were startled again when presented with a new pattern. This shows that birds habituate to one specific colour or pattern and don't simply generalize their responses to all conspicuous displays. Finally, when birds had habituated towards one combination of forewing and hindwing pattern, they then still showed a startle response to the same hindwing pattern but paired with a different (but still familiar) forewing appearance. Thus, anomaly can also cause birds to be startled.

In a third series of experiments in the early 1990s, Victoria Ingalls used the same apparatus as Vaughan to further investigate the importance of novelty and conspicuousness.[5] In this work, birds initially trained with grey discs were more hesitant towards coloured discs marked with contrasting black bands compared to discs of uniform colour. As such, while novelty matters, the startle effect is greater with enhanced conspicuousness, and this finding may reveal why many moth hindwings comprise both bright colours and banding patterns. Furthermore, birds were more startled by yellow and red colours than by blues, greens, or purples. On the one hand this is consistent with red and yellow being common in warning signals, but it's hard to know why the red and yellow colours worked so well. They might simply have been more conspicuous than blues and greens towards the avian visual system against the experimental background, so we can't be sure that this is an effect of colour itself. Finally, Ingalls found that when she presented birds with a greater diversity of coloured disc types the birds took longer to habituate than those presented with fewer colours, suggesting that diversity in hindwing appearance inhibits predators from learning to ignore or habituate to the displays.

In essence then, multiple aspects of startle displays seem important, including how conspicuous and visually contrasting the display is, how novel and rare it is, and elements of anomaly with unusual combinations of colours and patterns. This work also sheds light on why many animals with startle displays vary in coloration among species, since birds took longer to habituate to the startle displays when they encountered more colour types. For example, in the case of *Catocala* moths, if predators such as birds were to keep encountering the same moths they would eventually habituate to their specific display. Consequently, the startling effect would decline and the predators would learn to ignore the display and attack the moths instead. By each prey species evolving hindwings of different colour types, the chances of the same individual predator encountering the same colours are greatly reduced. As a result, individuals from each *Catocala* species stand to benefit because their respective colour type is made rare. Indeed, as many as thirty to forty species of *Catocala* can occur in the same locations, and interestingly coexisting species of *Catocala* tend to use different hindwing colours or patterns. One issue yet to be properly investigated in *Catocala* moths (and other analogous groups) is how the frequencies of hindwing colours vary in a given location over time. As we saw in Chapter 4, various camouflaged species are polymorphic, occurring in different forms of the same species. This may prevent predators from forming search images for the most common types, and over time the population should fluctuate, with more common morphs declining and less-common types increasing in frequency. This idea should apply among similar-looking species too. If the effectiveness of startle displays is based partly on unexpectedness and novelty, individuals from less-common species might have an advantage because when they are encountered the predator is unlikely to have seen them recently. We might, therefore, expect frequency-dependent selection and cycles in hindwing types over time. However, Sargent's work suggests that species with different hindwing patterns are relatively equal and stable in frequency over time, in at least some locations.[2] This might occur if predators simply base their expectations on only the most recent hindwing type they have encountered, rather than generally assessing how common different types seem to be over longer time periods. As yet, we don't know exactly what the true situation is, but it's interesting that the frequency of hindwing patterns and moth species is likely to be so strongly linked to aspects of predator learning and generalization behaviour.

Startle displays are by no means restricted to moths. They have also been studied in cuttlefish, and this research has helped us to understand when and how startle displays are used. Keri Langridge and colleagues at the University of Sussex showed that juvenile European cuttlefish (*Sepia officinalis*) use startle-like displays when faced with potential threats.[6] Cuttlefish, as we have already observed, are masters of camouflage and have the ability to change colour and pattern within seconds to match their background rapidly. But they also use their colour change skills for signalling to one another and towards other animals. Langridge presented cuttlefish with species that are potentially threatening to them, including crabs, juvenile dogfish sharks, and juvenile sea bass, and recorded how the cuttlefish responded towards these threats (without allowing them to be eaten). The crabs and dogfish primarily hunt with non-visual senses, whereas sea bass have vision effective enough to break the cuttlefish camouflage. When threatened by the dogfish and crabs the cuttlefish fled, but when threatened by the sea bass they displayed dark spots, raised their tentacles, and made their whole body look larger (Figure 41). Cuttlefish somehow, like many vertebrates, discriminate between different potential threats and choose a defence that is most effective against each one. Indeed, subsequently Langridge showed that cuttlefish do not use their

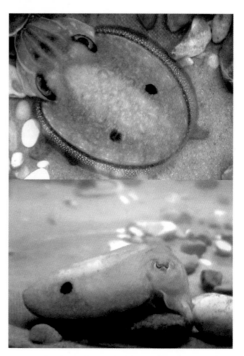

FIG. 41. The startle or threat display of a cuttlefish, making its body larger and displaying two prominent dark spots.

Images Keri Langridge

startle display against larger adult sea bass (instead they often flee or hide). This diverges from the behaviour of many insects, which use their displays against a dangerous and clear threat, because the cuttlefish actually do not use startle displays against the most serious predators. Instead, their display seems more useful against animals that are more of a nuisance than a major danger, or when smaller fish investigating them could attract the attention of larger dangerous adults observing from afar. Langridge suggests that this difference arises because moths and other insects are limited in their flight abilities to escape predators such as birds and so cannot flee as effectively, whereas cuttlefish can rapidly escape by expelling jets of water to propel them.

Although startle displays are most frequently considered in the visual sense, there is no reason why they should not evolve to exploit the other sensory modalities. In fact, several studies suggest that acoustic startle displays exist in insects. As we discussed earlier, many moths that are attacked by bats have evolved both ears to hear the echolocation calls of their predators hunting, and sound-producing organs to signal back to the bats that they are toxic. Another approach is to release a sudden, startling burst of sound when a bat attacks, and it seems that some moths do just this. For example, past experiments have involved scientists training bats to take mealworms from a platform, at which point they were presented with the sounds of clicking moths.[7] In response the bats did show startle responses, especially when the moth clicks occurred in the final phase of the bat attacks (i.e. close to prey capture). However, bats also quickly habituated to the displays, suggesting that they may be of value only when bats don't encounter them frequently, analogous to the *Catocala* moth hindwings. Interestingly, ultrasonic clicks in response to bats have been recorded in tiger beetles and may also function as startle displays.[8]

The best evidence so far, however, that acoustic startle displays exist and work comes from a recent study by Veronica Bura from Carleton

University and colleagues.[9] They followed up an intriguing discovery that caterpillars of the North American walnut sphinx moth (*Amorpha juglandis*) produce a burst of sound when picked up and squeezed. The team first showed that this 'whistling' occurs in sound frequencies spanning those that could be detected by humans and birds and on through to ultrasonic frequencies, and that the noise lasts for several seconds. Caterpillars, like other insects, effectively breathe through tiny holes in their body called spiracles, which allow air to enter special tubes and oxygen to diffuse into their blood. By applying latex to the different pairs of spiracles on the caterpillars, blocking the entrances, Bura found that the caterpillars produced sound by expelling air specifically from an enlarged eighth pair of spiracles. This is similar to the way in which the famous hissing cockroaches of Madagascar produce sounds, thought to resemble those of snakes, when picked up. Next, the team presented live caterpillars to three captive yellow warbler birds and filmed what happened. The birds, as expected, prepared to eat the larvae, but when they picked them up the caterpillars whistled in response, causing the birds to invariably drop them, sometimes flying away and diving for cover. In fact, the birds did not eat a single caterpillar, and despite often attempting to capture the caterpillars several times they didn't habituate in trials lasting up to sixteen minutes.

Acoustic startle displays also exist in adult Lepidoptera, as has been demonstrated in peacock butterflies (*Aglais io*), a species that we will also return to shortly. When disturbed, peacock butterflies flick open and close their wings, producing ultrasonic clicks by rubbing the wing veins together. This 'hissing' sound is apparently produced to deter rodents, many of which can hear ultrasound. Peacock butterflies often roost and overwinter in colonies in locations where they would be vulnerable to mice, rats, and other rodents that would find them easy prey. Martin Olofsson and colleagues at Stockholm University presented butterflies to mice in the lab to test whether the peacock's display deters rodents.[10] They manipulated the butterflies' wings so that they could either produce sounds or were silenced (but could otherwise display and move their wings as normal), and then assessed how mice responded to the two prey types. Mice were more likely to flee when the butterflies were able to produce sound, suggesting that the rodents were indeed startled by the display. Presumably, this effect would also be greater if several butterflies in an overwintering colony were simultaneously disturbed. The study is convincing, although at the moment what we don't know is whether mice were just scared by any unexpected noise, or if there is something inherently startling or off-putting about the specific peacock sounds. Perhaps, for example, the butterfly sounds are especially salient in terms of the frequencies that rodents can hear well.

Before moving on from startle displays, there is another aspect to the displays of many species that we have yet to consider—one that has provided much debate over a number of years and that also relates to our earlier exploration of Batesian mimicry. A quick look through a field guide to moths and butterflies reveals the presence of a common feature found on many tropical and temperate species: so-called eye-spots. These circular markings, found in pairs on either side of the body, come in a variety of sizes, colours, and complexity, and are striking in how often they arise among species. Some spots are large, complex, and found directly near the middle of the wings, whereas others are smaller, often occurring in rows along the wing edges. It's worth considering eyespots in some detail, because they have attracted the attention of many evolutionary biologists seeking to understand predator–prey interactions and deception. Eyespots likely have several functions, but one primary use is when combined with a startle display.

The first question we might ask is whether eyespots do deter pred-ators (we will come to the 'why' question shortly). The first experiment

FIG. 42. The colourful wings and eyespots of the peacock butterfly (*Aglais io*). Experiments have shown that their startle displays and eyespots are very effective in deterring avian attacks.

Image © lianem/123RF

to test this was conducted by David Blest in the 1950s on peacock butterflies,[11] which have large gaudy multicoloured eyespots on the top sides of their wings, yet when the wings are closed they look dull, a bit like a dead leaf (Figure 42). Peacock butterflies tend to rest with their wings closed (unless basking), but flash them open and continue to open and close the wings when danger is close. Blest rubbed off the wing scales of peacock butterflies to remove their eyespots and then presented them to birds, showing that birds were less likely to be startled by eyespotless displays than by butterflies with their eyespots intact. This was useful early work, but unfortunately Blest did not include any control for the effect of physically rubbing off the eyespots on the butterflies. His results could simply have arisen if those butter-flies were no longer motivated or able to display as effectively as individuals whose spots were left intact.

It was almost fifty years later that clear evidence finally showed that eyespots do genuinely startle birds. Adrian Vallin and colleagues at Stockholm University essentially repeated Blest's experiments but under much more effectively controlled conditions and without the problems of Blest's manipulations.[12] They presented peacock butterflies to captive blue tits in laboratory rooms and recorded their interactions, including how likely the butterflies were to be eaten. In some individuals Vallin painted over the eyespots with a permanent black marker, whereas other individuals were painted on a different region of the wing away from the eyespots (this acted as a control for the manipulation of the eyespots potentially affecting the butterfly display). The trials showed clearly that eyespots were strongly effective against the birds. In thirty-four trials with butterflies with their eyespots intact, just one individual was killed. In contrast, thirteen out of twenty butterflies with the eyespots removed were killed, demonstrating how valuable the spots are over and above the other display components. What's remarkable is that the experiments were conducted in a small room, roughly two metres wide, and the trials lasted a whole thirty minutes (unless a butterfly was eaten), yet just one butterfly with eyespots was killed across all trials. In many cases, birds repeatedly approached motionless butterflies yet were continuously deterred by the subsequent startle display. Thus, the effect was strong and prolonged. Interestingly, however, the team's manipulations of the butterfly wings to prevent them making the hissing sound did not affect the birds' willingness to attack, further suggesting that the acoustic component of the startle display is directed towards rodents rather than birds.

Vallin and colleagues then undertook another similar set of experiments whereby they presented both peacock butterflies and eyed hawk-moths (*Smerinthus ocellatus*) with and without eyespots to blue tits and great tits in aviaries.[13] Eyed hawk-moths are large moths with a pair of black and blue/white eyespots on their hindwings, surrounded by red coloration (Figure 43). Their display differs somewhat from that of the peacock, with the moths opening their wings when threatened and often rocking their body and wiggling around in the same place, keeping the eyespots visible throughout. The study again showed that eyespots work in deterring birds, but the display of the two species was

FIG. 43. The eyed hawk-moth (*Smerinthus ocellatus*), which when attacked opens its wings to reveal striking eyespots on its hindwings that prevent attacks from birds.

Image © Татьяна Серебрякова/123RF

not the same because birds were more wary of the peacock than the hawk-moth. It is hard to know, however, whether this stems from peacock butterflies having more spots than the eyed hawk-moths (four instead of two), different spot colours, or differences in display behaviour.

These studies and others clearly show that eyespots enhance the effectiveness of startle displays, but the question is: why do they work? The most widespread idea, and the one favoured by most studies on eyespot function, is that eyespots mimic the eyes of larger predators (as the name 'eyespot' would suggest). A small foraging bird attacking a moth may interpret a pair of eyespots suddenly appearing as the eyes of one of their own predators (such as a hawk or other raptor), causing them to take flight instead. This is perhaps one of the most long-standing ideas of protective coloration known, with natural historians giving short descriptions in the early nineteenth century suggesting eye mimicry by butterfly markings.[14] Certainly, by the mid-twentieth century the idea was prevalent. That remains the case today in both the scientific and popular literature. Open almost any book or paper that discusses eyespots and it will almost certainly state or suggest with varying degrees of certainty that eyespots mimic eyes. But if we delve deeper into this idea then things become much more controversial than many studies would have us believe.

The idea that eyespots mimic eyes has been a subject of interest to me ever since my PhD work. Initially the aim of my thesis was to investigate both how camouflage and eyespots work to prevent predation of animals. Although I ultimately focussed more on camouflage, my supervisors and I performed several experiments on eyespots and what makes them effective in scaring birds—experiments that I continued for several years thereafter.[15] I was always somewhat sceptical of the assumption that eyespots mimicked the eyes of larger animals. It's not that I didn't believe eyespots *could* mimic eyes, but rather that there was little actual evidence they did. Beyond this, I also was not especially convinced that some eyespots even looked like eyes. Take the peacock butterfly, the subject of Blest and Vallin's work. It is common for people to argue that its eyespots mimic eyes, but if you look at its wings the spots comprise a range of colours like blues and reds, plus various shapes and asymmetries that are not found in most eyes. The same can be said for the colours found on eyed hawk-moths. If they mimic eyes, why should they be so different to real eyes in arrangement and colour? More importantly, as I have tried to emphasize throughout this book, we should not in any case rely on what we perceive because it's the predator's vision and brain that matters. In addition, there were a number of other reasons to be cautious of the eye mimicry idea, because many of the arguments purporting to favour eye mimicry could be explained in other ways.[16] For example, many eyespots are circular in shape and comprise a series of rings around the central circle. In some Lepidoptera, the central area comprises a black region with a white spot or 'sparkle'. This is often assumed to mimic the dark pupil of an eye with a patch of light reflecting from it. Beyond that, there can sometimes be a lighter ring or colours, such as yellow, assumed to mimic an iris. This all sounds reasonable, but work by genetic and developmental biologists on butterflies has shown that circular shapes and concentric rings are relatively easy to develop during the process when a butterfly or moth forms in the pupa. Essentially what happens is that the eyespot starts to develop at a central focal point, from which chemicals diffuse outwards in a radial fashion, whereby cells that will become the wing scales respond in a particular manner to different concentrations to determine their ultimate colour. Because of this radial diffusion, it should be easier to develop circular-like shapes than other shapes. Likewise, many eyespots come in pairs, one on either side of the body, frequently assumed to mimic a pair of eyes. Yet, having a pair of features on either side of the body, termed bilateral symmetry, is extremely common in animals for numerous traits, simply due to the way that body plans work and because many animals are symmetrical about their midline. Finally, no experiment had ever actually directly tested eye mimicry against other competing explanations. The experiments by Blest, Vallin, and others convincingly showed that eyespots work to deter predators, but they do not show *why* eyespots work.

So if eyespots don't mimic eyes, how else might they work? There are several ideas that are not mutually exclusive. In the last couple of decades or so there has been a range of work showing that birds and other predators are often cautious of novel or unfamiliar food items, and that this avoidance is sometimes enhanced by stimuli with conspicuous or salient visual features, such as bright colours or strong markings. This causes predators to either directly avoid anything new (neophobia), or to be cautious before attacking anything that looks like it could be less than favourable (not dissimilar to how many humans stick to eating what they are familiar with and avoid trying certain other things). This makes sense because plenty of things in the environment are dangerous or unpleasant to eat, and so some degree of caution is likely to be favoured in foraging. Eyespots are often very conspicuous features—stimuli that catch the eye and attention—and their strong visual contrasts, and bright colours and patterns, should be good at stimulating the sensory cells in the visual systems of predators. These strong visual signals could therefore enhance avoidance behaviour in predators.

There is another potential reason why eyespots might prevent attack that primarily applies to startle displays. Work in humans and other animals has shown that there's a limit to how quickly our visual systems can process and evaluate visual information. Too much information and the time needed to process it is longer. This means that if you present someone with a stimulus of simple colours and patterns they may be able to respond faster and more appropriately than when presented with a stimulus of complex colours, shapes, and patterns. Eyespots, including many involved with startle displays, are clearly complex, comprising a range of shapes, colours, contrasts, and patterns, as well as sudden or repeated movements and sounds. In such cases, they may just be a very good way of overloading the sensory systems with a sudden stimulus, causing the predator to hesitate and allowing a prey animal to escape. In theory this could work for eyespots that are continuously on display too (some Lepidoptera have eyespots on their wings that are always exposed) because many predators are under severe time constraints and need to trade off making a fast and an accurate decision when hunting. Therefore, if they have to decide quickly whether to attack an object or fly away and search for something else, the complex nature of eyespots could cause the latter behaviour, favouring a fast response.

Over a period of five years or so, with various colleagues, I conducted experiments to distinguish between the eye mimicry theory and explanations that were based on eyespots working as they were highly conspicuous and visible.[15] This is not entirely straightforward because the theories need not be mutually exclusive. However, it's possible to make contrasting predictions. Specifically, we would expect predators to show greater avoidance of prey with eyespots that have a closer similarity to eyes if eye mimicry is important, whereas if being visually striking is what matters then we would expect eyespots to be most effective when highly contrasting and conspicuous (regardless of how closely they resemble eyes). To test these ideas we made artificial 'butterflies' from printed triangular pieces of waterproof paper with a dead mealworm, just as for the camouflaged moth experiments in Chapter 4. Again, our aim was not to mimic any real moth or butterfly species, but rather to make a prey item that birds would respond to as prey to test general principles of how eyespots might work. Targets with different patterns were pinned to trees in woodland and checked over several hours and days to determine if the target had been attacked and eaten by a bird. By printing eyespot patterns with different shapes, colours, and contrasts, we could alter the level of conspicuousness and eye-likeness of the targets and see how this affected survival.

One of the first experiments involved simple eyespots that comprised one spot (no concentric ring) that were either black, white, or a shade of light or dark grey (Figure 44). The birds did not differentiate between white or black spots (or dark and light grey), but they attacked targets with white and black spots less than those with grey spots. Because the white and black spots contrasted more strongly with the background 'wings' than the grey ones, this showed that contrast was an important factor in the effectiveness of the spots, with higher contrast spots avoided more. In addition, a single dark 'pupil' was not attacked less than a non-eye-like white spot, casting doubt on the eye mimicry idea. Next, we made spots that comprised black and white components: either with a white centre and black surround, or a black centre and white surround. These two stimuli types had exactly the same visual contrast and conspicuousness, but the latter one was more eye-like because it more closely resembled the dark pupil and light iris

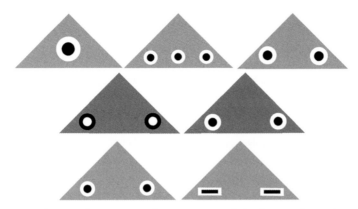

FIG. 44. Example paper butterflies used to test some of the features that make eyespots effective in scaring away birds. In experiments with wild birds, targets were less likely to be attacked when they had either one large or three small spots, than when possessing two small spots. In contrast, there was no difference in the survival of prey with spots with black or white centres, or with circular or rectangular shapes.

Images Martin Stevens

of eyes. Here, we found no difference in survival, again suggesting that the level of eye mimicry was unimportant. We also conducted experiments changing the shape of the spots, from circular to square and rectangle, and again found no difference in survival. Being circular did not improve the spots' deterrent effect. However, prey were avoided more when the spots were larger and greater in number, including having three as opposed to two spots. Overall then, this series of experiments (and others I have not mentioned) pointed clearly in favour of conspicuousness being an important determinant in eyespot success, but provided little if any support to the eye mimicry theory.[17]

More recently, a study by Ritwika Mukherjee and Ullasa Kodandaramaiah from the Indian Institute of Science Education and Research presented chickens with a choice of two paper butterfly models to attack (broadly mimicking the species *Junonia almana*).[18] They tested how likely chickens were to investigate and peck at either model when the butterflies were marked with a series of spot-like features. From several experiments, a number of key findings were made. Of these, only one result clearly supported the eye mimicry theory, in that chickens were less likely to attack a model with a pair of small eyespots (a pair of 'eyes') than they were to attack models with a single large spot (in contrast to previous work). They also showed no difference in attack between models with either a pair of small or large spots. However, partly consistent with both theories, chickens were more likely to attack models with a row of small spots than those with a pair of larger spots. Most interestingly, there was no difference in attack preference towards models with either a pair of normal (eye-like) eyespots versus spots that had been destroyed in shape into a pair of non-circular 'fan' shapes, which looked nothing like eyes. The authors were understandably open in that their work was not entirely consistent with either theory. What they do show is that more work is needed to understand these apparently conflicting results, and that the findings of studies can vary depending on the experimental approach used. Here, the work was based on a domesticated species in controlled laboratory conditions, presenting birds with two prey types at the same time to choose between, and where predator behaviour can be directly observed. This experimental design creates a simultaneous encounter where the predator is 'forced' to choose one prey or the other. In contrast, field studies present prey to a suite of wild predator species under more natural, but less controlled conditions. In these experiments, prey encounters are sequential, meaning that only one prey is encountered at any one time and the predator faces a choice not in terms of which prey item to attack of a pair, but rather whether to attack one prey item or leave it alone entirely. Neither of these approaches is right or wrong and both have respective merits, but such differences might explain some of the divergent findings.

Things so far are not looking good for the eye mimicry hypothesis. Yet, as so often occurs in science, other recent work does support this idea. To begin with, researchers at Heinrich-Heine-Universität in Germany performed an experiment very similar to our field experiments with fake paper butterflies marked with eyespots and varied the appearance of the markings.[19] This included changing their size and adding a small white highlight or 'sparkle' to different parts of the dark centre of the spot. Larger eyespots, and those with the white sparkle, made prey less likely to be eaten by birds. However, prey with the sparkle were attacked least when this was on the top part of the centre of the eyespot. This is potentially consistent with eye mimicry because in real eyes there would often be a white highlight where light bounces back off the top part of the eye because most natural light comes from above. Next, in 2015 a study by Sebastiano De Bona and colleagues at the University of Jyväskylä in Finland presented captive great tits (*Parus major*) with animated computer displays of stimuli with different eye-like features.[20] They had five treatments, with one resembling an owl butterfly (*Caligo martia*) with normal eye-like spots; a Neotropical species often used as a classic example of eyespots that resemble the eyes of owls with black 'pupils' and yellow 'iris' surrounds (Figure 45). They then created another treatment but with the spot arrangement reversed, now with a yellow centre and black surround, and another with the eyespots digitally removed entirely. Finally, they had two treatments showing the face of a pigmy owl (*Glaucidium passerinum*), one with the eyes open and another with the eyes closed. During the experiment, birds were released into the experimental arena and allowed to attack a mealworm larva next to the computer monitor. When they did, the blank white screen changed to display one of the treatment types. De Bona and colleagues predicted that if eyespots mimic eyes, then the great tits should be most scared of the owls with open eyes and the normal butterfly eyespots. In contrast, if

FIG. 45. The eyespots on the owl butterfly (*Caligo* spp.) have been shown to scare away birds by mimicking real predator eyes, presumably those of an owl.

Image Martin Stevens

conspicuousness matters then the eyespots with the inverted arrangement should be as equally aversive as the normal eyespots. The real eyespots caused more startle responses in the great tits than did the modified eyespots, and there was little difference in response by the birds to the real eyespots and the owl with open eyes. This shows that the eyespots did seem to be more effective through eye mimicry, and that they were similar in startling effect to real owl eyes. Why then did this study find clear consistent evidence in favour of eye mimicry when multiple other studies have not? There are important differences between this study and other work. De Bona and colleagues' experiment involved eyespots that were suddenly revealed in a startle display, rather than being continuously visible as in past work. In addition, the eyespots used were, to human eyes, much more convincing mimics than those used in previous studies, suggesting that aspects of conspicuousness may be important when the resemblance to eyes is imperfect, but not crucial when the resemblance is very strong.

The research undertaken on the owl butterfly spots is the best evidence so far that butterfly eyespots can work by mimicking eyes, but is has also coincided with studies investigating eyespots found on caterpillars. Naturalists have long argued that some species of caterpillar mimic snakes, both in terms of movement and in their coloration and posture. Many of these potential mimics have eyespots. Thomas Hossie, a scientist at Carleton University in Canada at the time, drew my attention to a quote from Bates in 1862 from his Amazonian travels:

> The most extraordinary instance of imitation I ever met with was that of a very large Caterpillar, which stretched itself from amidst the foliage of a tree which I was one day examining, and startled me by its resemblance to a small Snake. The first three segments behind the head were dilatable at the will of the insect, and had on each side a large black pupillated spot, which resembled the eye of the reptile: it was a poisonous or viperine species mimicked, and not an innocuous or colubrine Snake; this was proved by the imitation of keeled scales on the crown, which was produced by the recumbent feet, as the Caterpillar threw itself backwards...I carried off the Caterpillar, and alarmed every one in the village where I was then living, to whom I showed it.[21]

Hossie probably encountered the same species (either *Hemeroplanes ornatus* or *H. triptolemus*) in his own travels in the Peruvian Amazon, albeit around 150 years later (Figure 46). Like Bates, Hossie was convinced this was snake mimicry, but realized that this needed experimental testing and undertook a PhD on caterpillar snake mimicry with his supervisor Tom Sherratt. In the first of their studies,[22] Hossie and Sherratt tested whether the eyespots found on caterpillars do actually deter bird predators. They made fake model green caterpillars from pastry coloured with food dye, such that birds could attack and eat them (Figure 47). Some of the caterpillars they marked with eyespots at one end of the body. The models were pinned to the branches of trees and monitored for how many were attacked over a period of around four days. Eyespots did indeed reduce predation, but only on caterpillar

FIG. 46. A species of caterpillar from Costa Rica with a striking similarity to a venomous snake, most likely *Hemeroplanes ornatus*.

Image © S.J. Krasemann/Getty Images

FIG. 47. Final instar of the Canadian tiger swallowtail (*Papilio canadensis*) caterpillar (left), and an example of the artificial caterpillar prey with an eyespot that Hossie and Sherratt used to test the role of eyespots in potential snake mimicry (right).

Images Thomas Hossie

models that were also well camouflaged, as opposed to prey that were uniform in coloration. Next, they used caterpillars with and without eyespots, with the models being either cylindrical in shape or with the front end of the body widened, as occurs when some caterpillars display to a predator. Here, eyespots again provided an advantage against predators, as did the enlarged body section. These results show that birds were cautious of caterpillars that had more character-istics of snake mimicry but did not show whether eyespots worked by resembling eyes.

To address this issue, Hossie and Sherratt teamed up with John Skelhorn (the same person who studied caterpillar twig mimicry in Chapter 4).[23] They presented chicks in an aviary with artificial caterpil-lars that either had eyespots and the enlarged body sections or not. They also changed the location on the prey where the features were found by placing them either at the front of the body, as would be found in both snakes and snake-mimicking larvae, or in the middle of the body. For both the eyespots and the enlarged body sections, chicks were more wary of the displays when at the front of the body than when in the middle. Because the position of the display components on the body should not substantially change their conspicuousness, but should affect the accuracy of snake mimicry, the results suggest that birds avoided the prey because they feared a snake rather than simply avoiding something unfamiliar.

So there is good reason to believe that eyespots in caterpillars do mimic eyes. But there remains a general puzzle because many appar-ently snake-mimicking caterpillars live in places where no dangerous snakes actually exist. If there is no dangerous model for the predators to fear then how can mimicry work and be maintained? Hossie sug-gests a number of possibilities, including that some birds migrate from areas where dangerous snakes do occur, or that a general fear of snakes is innate or 'hardwired' in many birds from encounters that occurred deep in the evolutionary past. If this behaviour was inherited then it may still persist today in birds from areas without snakes. At the moment, we do not know the answer to this problem.

In general then, what at first seems to be a textbook example of mimicry and deception is less clear-cut than is often assumed. There have been approximately ten recent studies testing the eye mimicry hypothesis, with most having either failed to find evidence for eye mimicry or at best providing ambiguous results. However, a handful of recent studies have started to, at long last, support this idea. At the moment we need more work to understand why such contrasting results have been found, which must at least partly relate to specific features of the way experiments have been designed and conducted. What we can probably conclude so far is that eye mimicry may well be important for the function of the eyespots of some species, perhaps especially in caterpillars, but it also does not seem essential for all eyespots to work. Of all the work so far, perhaps only three studies have clearly supported the eye mimicry idea, and so it would be premature to conclude that *all* eyespots mimic eyes, or that conspicuousness is not important. Beyond this, eyespots tell us that we need to be cautious about relying too much on our own impressions of animal colour patterns and deception without objective experimental evidence, and that sometimes results can be more complex and intriguing than we might imagine.

As we move on from startle displays, it is worth considering other types of eyespot because they illustrate another route by which prey deceive predators and manipulate their attacks. The eyespots we discussed so far are normally large and centrally placed on the body or wings for maximum effect, but many Lepidopteran eyespots occur as small spots at the wing margins, often in rows (Figure 48). It seems that these eyespots do not work by scaring predators, but rather by deflecting attacks away from the vulnerable body and to the small

FIG. 48. A large wall brown (*Lasiommata maera*) with its row of small eyespots near the margins of the wings, which stand out from the otherwise camouflaged appearance of the butterfly.

Image Martin Olofsson

spots found at the outer margins instead. Because butterflies and moths can often still fly with considerable wing damage they could survive losing part of their wing and live to fight another day. While this seems to be a simple idea, it has been notoriously difficult to demonstrate that it actually works. Initial studies documented wild butterflies with triangular beak-shaped chunks missing from their wings where the eyespots would normally be. Such damage seems to be more common in species with marginal eyespots, consistent with a deflective function. The problem with these sorts of data, however, it that we cannot be sure that the beak marks are due to eyespots directing avian attacks towards the wing margins, or their presence drawing greater numbers of predator attacks to those species overall. It might just be that butterflies with eyespots are more likely to get attacked in the first place. What we need to do is to conduct experiments with predators and change the presence and features of eyespots to see if that influences attack behaviour. Indeed, several laboratory experiments did just this—they presented butterflies with eyespots to both birds and lizards and analysed where the predators attacked. Unfortunately, these studies showed very little evidence that predators were more likely to attack the eye-spotted wing margins of butterflies than elsewhere on the body.[24]

It was not until 2010 that direct evidence for deflection started to appear. Martin Olofsson and colleagues in Stockholm (many being from the same team studying peacock startle displays) presented mounted wings of the woodland brown butterfly (*Lopinga achine*) to captive blue tits.[25] The woodland brown is a species with rows of eyespots with black, white, and yellow rings on the undersides of its wings. The white markings on the eyespots also strongly reflect UV light. Olofsson and the team tested the blue tits' responses under three light conditions: low light levels without UV, and high and low light levels with UV light added. Under the high light and dim light lacking UV the birds generally attacked the head region, but under the low light with UV added they switched and now attacked the eyespot regions instead (Figure 49). This was the first evidence showing unambiguously that eyespots can, in principle, deflect avian attacks. However, the broad relevance of these findings is unclear for two reasons. First, the illumination treatment where the effect was observed, low light with added UV, was somewhat artificial and had much higher levels of UV than would normally occur in nature, which may have made the white spots artificially bright. Second, running alongside the row of spots on the butterfly wing is a prominent white stripe. This would also glow strongly under UV and so it's hard to be sure that the birds were

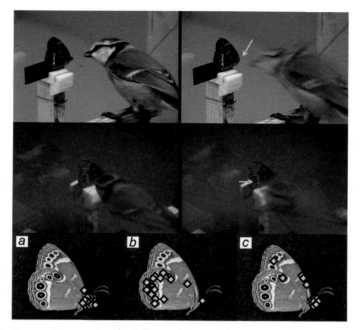

FIG. 49. An experiment testing the deflective function of eyespots. Such investigations have assessed whether eyespots cause predators such as birds to attack the regions of the wings where the spots are found. Here birds attacked the head region of the butterfly under high light levels (top), but the eyespot areas of the wings under dim conditions with enhanced UV light levels (which birds can see). Only under dim light with enhanced UV did birds attack towards the spots (b), whereas under high light levels with UV or dim light levels without UV they attacked the head (a, c).

Images Martin Olofsson

directed towards the spots specifically, the stripe, or the general area because that was the only part of the butterfly that was visible (few white markings occur towards the front of the butterfly).

More recently, in 2013, a further experiment by Olofsson and colleagues validated their earlier work by demonstrating that eyespots deflect the attacks of birds against butterflies seen on natural backgrounds and under more natural lighting conditions.[26] They presented blue tits in the lab with dead butterflies, with and without a prominent eyespot, against a background of tree bark. Here, some birds did misdirect their attacks towards the hind regions of the butterflies when eyespots were present. Overall, nine out of forty-seven birds (19 per cent) faced with a butterfly with an eyespot attacked the eyespot region of the wings, whereas just one of forty-one birds (2 per cent) attacked the hind margins when no spot was present. This tells us that deflection can work and probably well enough for selection to favour them, but perhaps unsurprisingly given past work, the effect is not dramatic. Interestingly, deflection was more likely to occur when birds attacked quickly, whereas birds that hesitated before striking were less likely to go for the spot. As such, eyespots are more likely to deflect an attack when the predator is pressed for time. Beyond this, most work on eyespots and deflection in butterflies has focussed on vertebrate predators, but very recent work has found strong evidence that preying mantis attacks are also directed towards eyespots, leading to higher escape behaviour in butterflies with than without spots.[27] Therefore, perhaps part of the difficulty in demonstrating deflection in past work has arisen through not considering other important predator species.

A more elaborate version of the deflection idea in butterflies is the presence of 'false heads'. Many butterflies, especially those belonging to the Lycaenidae (which we met in Chapter 1, with some species tricking ants into raising their larvae), have elaborate structures on the hind part of their wings that resemble a fake head (Figure 50). This suggestion originated at least 200 years ago, and was highlighted again by Poulton in 1890. Sometimes, the bluff can be really quite convincing. For example, the chocolate royal butterfly (Remelana jangala) is a beautiful species with males having bright blue iridescent wings on the top surface and a deep chocolate brown on the underside. When the butterfly sits with its wings closed, the hind parts of the wings show off a pair of black spots surrounded by a silvery colour, and from the backs of the wings come four thin extensions with black and white markings, which seem to

FIG. 50. A branded imperial (*Eooxylides tharis distanti*) butterfly from Singapore with long tail-like extensions to its hindwings that may work to create the impression of antennae and a false head that help deflect predator attacks.

Image Martin Stevens

have evolved to mimic the structure and colour of the front two legs and antennae of the butterfly. Some lycaenids, such as the red-edge blue (*Semanga superba deliciosa*) accompany these structures with apparent behavioural mimicry, 'scissoring' their wings together back and forth making the false legs and antennae move in a manner like the real ones. The idea of a false head is again that it can cause the predator to direct its attack away from the head region and towards the back of the butterfly. In fact, if it works well the butterfly might fly off at the same time, such that the predator misses the back end of the butterfly entirely. Sadly, although this phenomenon is ripe for testing and a wonderful potential example of deception and mimicry, very little work has directly investigated it.[28]

While deflection of predator attacks has most often been discussed with regard to butterflies, there is no reason why it should be restricted to this group of animals. In fact, the defence has probably also evolved in fish, tadpoles, beetles, and lizards. For example, some amphibian tadpoles have coloured tail fins or dark markings that may protect them from predatory dragonfly larvae. Experimental work painting dark spots on the tails of tadpoles results in dragonfly larvae attacking the tails instead of the body, increasing the chance of escape because the tail can tear easily and break away. Indeed, tadpoles with spots on their tails may be two to three times more likely to escape an attack than those without tail spots.[29]

Many fish, especially tropical ones, also have dark eyespots on the rear ends of their body, often on the dorsal fin. These could misdirect predatory attacks to an area from where the fish is swimming away. Experimental tests that this does happen are rare, but one recent study demonstrates nicely how eyespot development can be mediated in individuals based on perceived predation risks. Oona Lönnstedt at James Cook University in Australia, with several colleagues, studied how the eyespots of juvenile ambon damselfish (*Pomacentrus amboinensis*) develop in response to predation risk.[30] Juveniles are yellow in colour, having a black eyespot with a white surrounding ring on their rear dorsal fin either side of their body. These spots tend to fade away as individuals grow and mature. The team presented damselfish with the sight and smell of either a predatory fish—a dusky dottyback (*Pseudochromis fuscus*), which as we know from Chapter 2 is a common predator of small damselfish where they occur in the Indo-Pacific—or a benign control fish of similar shape and size to the predatory fish (a goby), or simply no visual or chemical cues of another fish at all. The goby was included to test how the cues of another species that is not a threat might cause any changes in eyespot development. Over a period of six weeks, the juvenile damselfish that grew up under heightened predation risk developed larger eyespots than those under low perceived risk. They also developed smaller real eyes when exposed to predators. This suggests that the function of the spots is to confuse predators about which end of the fish is the head and perhaps to deflect attacks because the spots become more detectable than the eyes when predation is more likely. Next, the team released the fish on to a coral reef and measured their continued presence (as a proxy for their survival against predators) through scuba-diving. When released, fish previously exposed to predators had only 10 per cent mortality after seventy-two hours, compared to 60 per cent mortality in fish that had not been exposed to predators. The implication here is that differences in eyespot size, as well as other changes in behaviour (such as increased hiding and vigilance), protected those fish from predators in the wild. Therefore, in some animals the exact features of eyespots and deceptive stimuli can be modulated during development depending on the risk of predation.

More recently, work by Jesse Barber, now at Boise State University, along with several colleagues, has shown that deflection of predatory attacks can also occur outside the visual sense.[31] They studied how luna moths (Actias luna) were captured by bats in staged predation encounters, filmed in the dark with high-speed infrared cameras. Luna moths, along with some of their relatives, have elaborate 'tails' extending from their hindwings that spin when the moths fly. Barber and the team found that bats, hunting by echolocation, directed their attacks towards these tails in over 50 per cent of the trials. This significantly increased the likelihood that a normal moth would escape compared to individuals with their tails removed; bats were almost nine times more likely to capture moths without tails than those with them. Barber and colleagues also showed that long hindwing tails have evolved four times independently in the group to which luna moths belong (Saturniidae), providing tantalizing evidence that deflection of predator attacks may be important beyond visual traits. However, quite how this defence works is as yet unclear.

Let's continue our discussion of bats and moths, because they illustrate another way that prey animals could interfere with a predator's ability to capture prey and finish off an attack: sonar jamming. As we know already, many moths produce ultrasonic signals towards bats, sometimes to warn them that they are toxic and sometimes to startle them into aborting an attack. There is yet another method they could use which directly disrupts or jams the echolocation equipment bats have, preventing them from being able to target prey animals accurately. The idea behind 'sonar jamming' is that an insect could release a burst of ultrasonic sound when attacked, which could disrupt the bat's ability to process the returning echoes from its calls correctly, or create some other sort of misleading information. The most likely candidates for this tactic are again certain species of tiger moth, which are well known to emit bursts of ultrasonic sounds.

Initial work on sonar jamming was tantalizing, albeit not clear-cut.[32] First, studies showed that some moths produced bursts of sound in the terminal phase of a bat attack. This is something that would be predicted by the jamming hypothesis but not if the sounds act as a warning signal, because the warning sounds should occur well in advance to warn that a moth is toxic and worth avoiding altogether. In addition, when bats were presented with bursts of sound just prior to when returning echoes should arrive, this interfered with distance estimates made by the bats, rather than causing the bats to abort their attack, as would have been expected with a startle response. Next, towards the end of the 1990s recordings were made from bat nerve cells that are known to function in judging time delays and distance estimates to objects, showing that moth clicks could interfere with these neural responses.

Sometimes, interference clicks could delay the actual occurrence of nerve signals that would normally encode a target when they were timed correctly. All of this was broadly consistent with the theory that bat echolocation calls could be interfered with by appropriately timed sounds, and that moths could adopt this approach successfully. But clear evidence was hard to obtain, partly because of the speed of the interactions between bats and moths, but also because the sounds were occurring outside the natural hearing range of human observers.

Recently, however, clear evidence has been found thanks to Aaron Corcoran and his adviser William Conner and colleagues at Wake Forest University in the US.[33] Conner's lab was well placed to test this theory, using bats presented with moths in a 'bat cave': a dark room with walls lined with acoustically controlled foam to mask sounds from elsewhere. The team also had infrared high-speed video cameras, so that in addition to recording the sounds that bats and moths made, they could film in great detail and in pitch dark the behaviour of the bats and how they approached and attacked the moths (Figure 51). Their first study established that a species of moth, *Bertholdia trigona*, could effectively jam the echolocation calls of big brown bats (*Eptesicus fuscus*). *B. trigona* is a type of tiger moth from the south-western US. It's a pretty moth, with a silvery brown body and wings, edged with thin red lines and patches of orange along with two yellow blotches at the edge of either wing that look as if they could function as disruptive camouflage. It is not toxic, and bats eat this species perfectly happily when given the chance. Individual moths also have sound-producing organs on their thorax called tymbals, which we came across in Chapter 5. They distort the tymbals with muscles to produce bursts of ultrasonic sound. In the experiments, moths were tethered to a line hanging from the ceiling in the middle of the room (to prevent them landing or flying away), and bats were allowed to hunt them over a sequence of seven nights. Whereas for startle displays we

FIG. 51. A Townsend's big-eared bat (*Corynorhinus townsendii*) attacking the moth *Bertholdia trigona*.

Image Aaron Corcoran

would predict that bats should habituate to the moth sounds relatively quickly, sonar jamming should work largely regardless of bat experience because of the way it directly interferes with echolocation calls. Consistent with this, bats struggled to capture the tiger moths compared to non-clicking control moth species, and they did not improve with time and experience. Furthermore, the bats seemingly tried to adjust the timing and frequency of their echolocation calls to reduce interference from the moth clicks. The final proof was that when the team damaged the moth tymbals, so that they could no longer produce sounds, the bats had no trouble attacking and eating them.

The team next turned their attention to why sonar jamming works, through further experiments with big brown bats and *B. trigona* in the experimental room. Bats were again poor at attacking the moths, capturing them in just one-third of the time even towards the end of the seventh night. However, there was no evidence that bats made misdirected attacks at entirely different locations where a moth did not exist; that is, there was no evidence that the moth clicks gave misleading information that they were found elsewhere. Instead, the bats often attacked the general area where the moth was found but missed their precise location, sometimes by around fifteen centimetres or more. This is consistent with the moth clicks interfering with bats' assessments of distance, just as the neurobiological studies some years earlier suggested. So sonar jamming does seem to work through deceiving bats about the moths' precise location, though at present it is not clear how widespread it is, but preliminary work suggests this tactic may be found in various tiger moth species and it probably occurs elsewhere too. In some regards, sonar jamming is a bit like the idea of motion dazzle that we discussed in Chapter 4, as both interfere with a predator's ability to locate precisely the position of a prey animal and capture it. Both tactics seem to disrupt the sensory system (vision or hearing) from accurately encoding information about the location of an object.

We will finish this chapter with another type of bluff and distraction used to deceive predators. Various birds belonging to several families use distraction behaviours, including the so-called broken wing display.[34] For example, when faced with an approaching predator a nesting bird will conspicuously run away while holding one of its wings limp, creating the impression of a broken or damaged wing. If successful, the predator follows the bird thinking it's an easy meal. Gradually, the predator is led further and further away from the nest, and when at a safe distance, the completely uninjured bird flies off and back to the nest. Such displays are most common in ground-nesting species, including many shore birds like plovers, when faced with mammalian predators such as foxes. There is variation in the form the behaviours take among species, and even across individuals of the same species; some birds accompany the display with erratic attempts at flight, as if desperately trying to escape, whereas others are said to run off in a crouched position resembling a small rodent and even squeaking periodically (though quite why they should do this instead of feigning injury is not clear).

Although the assumption behind distraction as a strategy is that the predator must be capable of interpreting the display and (mis)under-standing that it means an easy meal, this is not unreasonable given that many predators do specifically target injured animals as easier prey. Certainly, to humans at least, the displays can be convincing and it's hard to think of other explanations for what's going on, other than the bird perhaps simply making itself more conspicuous and attracting the predator's attention away from the nest, without any interpretation of ease of capture or injury on the predator's behalf. Interestingly, some birds performing these displays also seem to modulate how they are used depending on the level of threat. For example, a parent bird may cease the display if the predator loses interest or wanders off, and they will often not bother at all if the predator is walking on a trajectory that would take them harmlessly past the nest at a safe distance. This all suggests again that the behaviour is driven directly to draw predators away from the nest to protect the young or eggs. It is not always successful, however—there are accounts of the displaying bird itself being caught and eaten by predators, so the defence comes with a risk.

So what factors lead to the evolution of distraction displays? In the first instance, the environment in which the nest is found needs to be relatively open, such that the bird can see the predator coming from afar and fly away in good time. If it flies off too late then the very act of leaving the nest could give the predator a valuable cue as to where the nest actually is. In addition, the nest, eggs, and chicks need to be well camou-flaged and hidden from the predator. If they are not, the display would have little value because the predator would easily find the nest anyway. Furthermore, the principal predators should be mammalian or reptilian and not avian, because reptiles and mammals would approach from the side with a limited view of the nest area, whereas aerial predators would have a good perspective of the nesting area from afar. Beyond all this, an unfortunate aspect concerning distraction displays is that there are many anecdotal accounts of them but almost no proper experimental research. It is another subject ripe for study.

A somewhat similar but more extreme case of faking injury is a type of defence often called tonic immobility. Here again, the phenomenon is largely anecdotal, with only a limited number of experiments showing that some animals, including birds, mammals, amphibians, and reptiles, will freeze when captured by a predator. Chickens, for example, often show such behaviour when grabbed or even stroked, especially when being stared at or cornered. Some insects and other invertebrates perform tonic immobility too. In extreme cases, the behaviour manifests itself as death feigning: the animal seemingly pretends to be dead, including by protruding the tongue and bulging the eyes. Opossums do this while lying on their side and waiting for the threat to pass.

While at first death feigning instead of just trying to escape may seem an odd idea, there may be good reasons for why it works. The limited work that has been done suggests that predators, like domestic cats attacking birds, are more likely to prematurely halt an attack when an animal feigns death than when it continues moving, partly because a struggling prey animal often seems to heighten the attack behaviour of many predators. The unresolved question, however, is exactly why it works. One could argue that the prey animal is simply so terrified that it just cannot move, but this is not a very satisfactory answer and the fact that predators do leave prey alone suggests that it is an adaptive response. Two main possibilities present themselves. First, if there are multiple prey in an area it may pay a predator to kill one animal quickly and then leave it for a while to go after further rewards. This may happen once the prey animal is subdued, and so the prey being motionless when relatively unharmed may cause the predator to leave it alone too quickly. Some evidence, over fifty years old now, shows that ducks can survive fox attacks through immobility, especially against relatively naive foxes. Second, many predators want a fresh meal; old meat could be diseased, rotten, or just taste bad and be worth avoiding. So, prey could dupe a predator into thinking it's already dead, and hence leave it alone. This is less likely to work for a freshly caught animal that simply freezes, but could work for animals that quickly play dead when a predator comes close before it attacks.

With the end of this chapter we conclude our look at the many ways in which prey animals deceive predators. As we have seen, such approaches can be highly varied and work through many different mechanisms. Camouflage and Batesian mimicry prevent the predator from attacking in the first place. In the case of camouflage it either prevents an animal being detected at all, or causes the predator to misclassify the prey animal as some other uninteresting aspect of the environment, such as a dead leaf. These defences do not just require specific aspects of coloration to work, but also frequently involve behavioural adaptations and even the ability to change colour as well. Batesian mimicry also works by causing a predator to misidentify a prey animal as the wrong species. However, in this instance the object being mimicked is not uninteresting to the predator, but rather something to avoid altogether (such as a wasp). Frequently, however, these defences fail to work and the predator still mounts an attack. To combat this numerous species have evolved secondary defences that either stop the attack in its tracks, such as startle displays, or prevent it from being successful, as is the case with deflective markings, sonar jamming, and motion dazzle. Secondary defences can confuse the predator, surprise it, or overload it with so much sensory information that the prey animal can escape or the predator attacks in the wrong place. Alternatively, defences can directly mislead an aggressor into following a strategy that is flawed (such as trying to capture a seemingly injured but actually healthy bird), or even by suggesting that the predator itself may be at risk, as may be the case with eyespots. These varied approaches have led to incredible diversity among species of animals in appearance and behaviour, sometimes even driving differences in appearance within species. In addition, while much of what we know about defensive strategies by animals comes from the visual sense and coloration, a suite of studies is showing us that many analogues exist in other sensory modalities too, especially sound and hearing.

7

AN IMPOSTER IN THE NEST

· · · · · · · · · · · ·

Back in 2010 I was fortunate enough to visit a colleague, Keita Tanaka from Rikkyo University in Japan, and to see a remarkable bird that lives on the slopes of the iconic Mount Fuji. The mountain looks like a perfect volcano—a clear cone rising above everything around it, topped with snow for much of the year. At ground level the Japanese summer is hot and very humid, but at 2,000 metres, where Keita finds his study species, the air is cool and fresh. The forest that covers much of the mountain here is thick and dark, with the slopes steep and difficult to climb owing to the loose volcanic soil and rocks. We were searching for the nests of the red-flanked bluetail (*Tarsiger cyanurus*), a beautiful blue bird whose nests are well hidden underneath overhangs of dirt and vegetation. It was not the bluetail chicks we were looking for though, but rather an imposter that is sometimes found in their nests: the Horsfield's hawk-cuckoo (*Cuculus fugax*). After several days searching, we finally found what we had been seeking—a cuckoo chick. It sat patiently while it was weighed, photographed, and measured. Before returning it to the nest, we took a look at the undersides of its wings, which are the reason it's so special. They are a vivid yellow, almost glowing in the gloomy conditions, and are totally unlike the wings of any other bird known.

The reason why Horsfield's hawk-cuckoo chicks possess yellow wing patches had been elegantly demonstrated by Keita and his colleagues a few years earlier.[1] The cuckoo is a type of brood parasite. Females lay their single eggs in the nest of another species, so that those hosts or foster parents rear the cuckoo's young instead. When it hatches, the cuckoo chick throws out ('evicts') all the host's own young from the nest and thus monopolizes the parental care. At least, that's the idea, but it's actually not so simple because when host parents are presented with a single chick in their nest, instead of a whole brood, they often bring less food. The reason is simple: there's no point bringing enough food for three or four chicks if there's only one to feed. The parents might as well save some effort for potential future broods or other tasks. This creates a problem for the cuckoo chick because it is often much larger than the host chicks, and therefore needs a lot of food. Furthermore, the cuckoo is not at all related to the parents (being a different species entirely), and so it should try to manipulate the parents to bring as much food as possible. While the chicks of many birds try to extract substantial care from their parents they are often not entirely selfish because they share many genes with both their current siblings and potential future broods. That means there should come a point when a chick is sufficiently well fed that, from an evolutionary perspective, it should stop begging and 'allow' the parents to allocate investment to its siblings because the chick would pass on more of its genes this way. Keita showed that when host parents bring food to a hawk-cuckoo chick, it raises one wing and shakes it, flashing the bright yellow wing patch (Figure 52). As you may have guessed, this looks somewhat like the yellow mouth of a hungry chick (admittedly not an especially close resemblance, but remember that the nest and habitat are very dark). Importantly, Keita showed that the display does trick the parents; painting the yellow wing patches of the cuckoo chicks black (like the rest of the wings) causes the hosts to bring

FIG. 52. The Japanese Horsfield's hawk-cuckoo (*Cuculus fugax*), which raises a bright yellow wing patch when host parents come to feed it to induce them to bring more food.

Left image Martin Stevens; right image Keita Takanka

less food than to chicks with unaltered wing patches, or ones whose wings had been painted with a transparent solution. The parents even sometimes try to put food into the wing patches rather than the mouth.

The effort to reproduce in nature is considerable for many species, and so there exist a variety of tricks to minimize these costs and pass them on to others. This chapter and Chapter 8 are about how deception is used in reproduction, and how it works. The wonderful trickery of the Horsfield's hawk-cuckoo is but a drop in the ocean because the bird world is full of similar cheats. In fact, of the approximately 10,000 avian species, 1 per cent (or one hundred species) are brood parasites that use other species to rear their young.[2] They are found on every continent except Antarctica, and this particular way of life seems to have evolved seven times independently. The word 'cuckoo' is often used as synonymous with brood parasitism, but strictly speaking this is not correct because cuckoos are a particular family of birds, with some species being parasites but others not. In fact, approximately 60 per cent of cuckoo species actually raise their own young. In addition, brood parasitism has also evolved in birds that are not cuckoos. Overall, brood parasites come from a somewhat eclectic range of birds, including cowbirds from the Americas (five species), several groups of cuckoos found in South America, Europe and Asia, Africa, and

Australia (fifty-seven species), a number of honeyguides (seventeen species) and parasitic finches (twenty species) from Africa, and even one species of South American duck (the black-headed duck). The way they interact with their host species is also varied. Some brood parasites target just one or two species of host, whereas others use multiple host species, although often with a high degree of specialism towards each one. As we will discover in this chapter, brood parasites have come up with a diverse range of tricks to achieve their goal. Beyond birds, the act of getting another species to rear the young is encountered in other animals too, especially insects. We have already encountered the blue butterflies that trick ants into looking after their caterpillars (Chapter 1), and a variety of social insects employ similar approaches to reproduction. As we will discuss shortly, how they achieve this sometimes parallels their avian counterparts.

It's not hard to understand why a species would evolve to become a brood parasite; rearing chicks is both time consuming and involves considerable investment in things like building a nest, bringing food to the chicks, and defending them against threats. Many tasks involved with parental care, such as increased foraging, also expose the adults to height-ened risk of predation. So getting someone else to do all the hard work makes perfect sense. Furthermore, birds are for one reason or another constrained by how many eggs they can look after in a single breeding season, for example owing to limitations in nest size or the number of eggs they can physically incubate. Parasitic species on the other hand 'only' need to find enough host nests to lay their eggs and leave them.

Early work in the 1920s by Edgar Chance, a wealthy businessman and ornithologist, established the fact that common cuckoo (*Cuculus canorus*) females laid their eggs directly into host nests, rather than as once thought laying them on the ground and carrying them to the nest. He spent much time observing cuckoos in Britain through hides, and even made a pioneering natural history film about them in 1922. Chance also showed that some female cuckoos could, remarkably, lay twenty-five or more eggs in a single breeding season. Each egg is laid in a different nest, meaning that even if some host nests fail through weather or predation at least some chicks elsewhere may survive. Parasitism, therefore, sounds like a wonderful strategy—one that involves minimum effort for maximum output. So why are brood parasites not much more common? Among other things, many hosts fight back and do not simply passively accept foreign young in their nest. Brood parasites like common cuckoos are often very costly to hosts because they require the foster parents to expend substantial effort rearing a foreign chick, not to mention that many brood parasites kill most or all of the host's own young in the process too. So, it usually (but not always) pays for hosts to fight back. There's another reason, related to this, why parasites are not more common: they are more likely to go extinct than non-parasitic species. Studies of cuckoo evolutionary pathways have found that brood parasites tend to form new species faster than those birds with parental care. This probably reflects the initial strong advantages brood parasites have when targeting naive species without defences if new parasitic species exploit new hosts. However, the flip side is that as parasitic species become increasingly specialized, so do their hosts. If host defences become very good then the parasite may no longer be able to successfully exploit its host, and will eventually go extinct because it cannot rear its own young any more either. Thus, the current number of brood parasites may partly reflect the balance between high speciation and extinction.[3]

A prevailing feature of much work on brood parasites is the focus on the evolutionary arms race with their hosts. Imagine a species of cuckoo that starts to lay its eggs in the nest of a new, previously unutilized host. At first, the host would be ignorant of the cuckoo, having not suffered parasitism in the past, and there would have been no selection to look out for foreign eggs. Thus, the parasite gets a free ride. However, over time hosts should evolve defences against the cuckoo, one of the most common being the ability to spot a parasitic egg in the nest and throw it out (to reject it). Next, selection would favour any cuckoos that evolve eggs that mimic the colour and pattern of the hosts' eggs to prevent them being detected, leading subsequently to hosts with further defences (such as refined rejection behaviour), and so on. This type of interaction is referred to as co-evolution because it involves reciprocal changes in each party (cuckoo and host) brought about by changes in the other species. Co-evolution can be an extremely powerful process in generating diversity in nature, although it is by no means present in all parasite and host interactions.

Parasites not only need to get their eggs into the nest, but also to be raised by the hosts. As such, the parasite's job is two-fold. First, it must circumvent any defences the host may have in order to have their egg accepted, and then the chick should extract as much parental care as possible to grow fast and large. Richard Dawkins and John Krebs, two pioneers of the field of 'behavioural ecology' (the study of the costs and benefits of behaviour and how they evolve) called these stages deception and exploitation. These two processes are central to how brood parasites use deception to exploit hosts, and indeed the ways that hosts can fight back, so let's consider each in turn.

The most intensively studied of all the brood parasites is the common cuckoo, especially in European populations (Figure 53). Many of us associate the sound of the cuckoo with spring and early summer, as individuals arrive after their migrations from Africa to breeding grounds. The common cuckoo widely exploits more than ten hosts in Europe—it's hard to give an exact number because some hosts are used much more than others. In the UK, the most important hosts are the European robin, dunnock, meadow pipit, pied wagtail, and reed warbler, among others. Nick Davies, another pioneer of behavioural ecology, and colleagues at the University of Cambridge have been studying the common cuckoo in the UK for several decades. Their work has revealed a great deal about how parasites exploit hosts, and how hosts fight back. Davies' principal study

FIG. 53. The common cuckoo (*Cuculus canorus*) is one of the most widely studied species of brood parasites. Here, a single cuckoo egg (the large one) is found in the nest of a reed warbler. The cuckoo chick hatches first and evicts all the host eggs from the nest. As the cuckoo grows it often becomes much larger than the hosts, even the host adults, and begs vociferously.

Images Nick Davies

area has been Wicken Fen in Cambridgeshire, an idyllic region of the remarkably flat fenlands, with little riverways and plenty of reed beds for reed warblers to nest. Being a dull brown-grey, reed warblers may not be among the world's most exciting birds, but they have evolved a suite of defences against cuckoos, and, in turn, the cuckoos have a plethora of strategies to get around them. In the first instance, cuckoos are elusive birds, lurking in nearby trees while they scope out the area and evaluate potential nests to target. This helps them to time their laying in line with the host clutch, but secrecy also reduces the chances that the host might see them. This matters because host parents, including reed warblers, are more likely to show heightened defences against cuckoos when they have recently seen a cuckoo, or when they perceive the overall risk of parasitism in the population to be higher (i.e. when there are many cuckoos around).[4]

Once a cuckoo decides to lay an egg in a chosen nest, she swoops down and picks out one of the host eggs, quickly replacing it with her own egg (taking from just ten seconds to a minute or two). Then she simply flies away, her job done. But cuckoos often get caught in the act by hosts, and parents mob and try to drive them off. Davies, along with

Justin Welbergen, showed that reed warblers discriminate between different types of threat and adjust their behaviour accordingly.[5] By showing stuffed mounted birds to nesting reed warbler parents they found that warblers were more likely to physically mob and show alarm calling towards a cuckoo than a sparrowhawk or non-threatening bird (a teal). This behaviour also draws neighbouring birds to join in, with the host responses greater when they perceive rates or risk of parasitism to be higher. A difference in host responses makes sense because the cuckoo is a threat to reproduction but not to the adults themselves, and so the warblers should drive it off. In contrast, the teal is no threat and so can be ignored, whereas the sparrowhawk is a dangerous predator that should be avoided.

Cuckoos actually seem to exploit the fact that many host parents avoid mobbing sparrowhawks to circumvent host defences. At least as early as 1867, and again in his classic 1889 book, *Darwinism*, Wallace suggested that some cuckoos mimic birds of prey like hawks.[6] It's not hard to see why he thought this because many, including the common cuckoo, have characteristic hawk-like barred breast plumage, an elongated body, bright yellow eyes, and even the appearance in flight of a hawk. In fact, during Aristotle's time people sometimes thought that the cuckoos' disappearance from Europe during the winter (when they migrate to Africa) was because they transformed into hawks, something Aristotle dismissed because cuckoos lack talons and a hooked beak.

Davies and Welbergen tested the idea that cuckoos mimic hawks, again using mounted birds of different appearance.[7] In the first instance, they presented cuckoos, sparrowhawks, doves, and teals to blue tits and great tits at feeding tables. If cuckoos do mimic hawks, then the tits should show similar types of alarm and avoidance behaviour towards the cuckoo and the hawk, but they should be unfazed by the teal. This was exactly what they found: feeding birds showed no response to the teal, but alarm behaviours such as erecting their head feathers, making alarm calls, and reduced attendance at the feeders when faced with both the hawk and cuckoo. Next, Davies and Welbergen manipulated the appearance of models by wrapping their undersides with pieces of silk. To remove the presence of bars they added plain white silk to the breasts of the cuckoos and hawks, and to add bars to the dove they marked the silk with black lines with a felt-tip pen. Now, the tits showed reduced avoidance of unbarred birds compared to barred birds. However, this partly depended on the model species. Adding bars to doves initially worked but the tits soon learned to ignore it, whereas removing bars from hawks had relatively little effect. The effect of barring on cuckoos was, however, more marked. Essentially, with bars the birds treated cuckoos as hawks, whereas without bars the tits responded as if the cuckoos were doves.

Cuckoos are no threat to great tits or blue tits because neither species is used as hosts, nor should they be familiar with cuckoos. It's therefore hard to interpret the above study in any way other than the tits categorizing cuckoos as hawks, and that the barring is an important component of this deception. It does not tell us though how actual cuckoo hosts respond, or whether hawk mimicry gives cuckoos greater access to nests. To answer this question, Welbergen and Davies presented reed warblers in the fens with models of cuckoos and hawks, again with their barred plumage manipulated. Like the tits, reed warblers tended to avoid hawks, largely ignored doves, but mobbed cuckoos. Again, barring was shown to be important because warblers stayed further away from barred cuckoos than those without bars, and mobbed unbarred cuckoos more intensely than barred ones. All this should mean that real cuckoos could gain additional opportunities to access host nests through deception.

Circumventing initial host defences is just the beginning for many brood parasites because hosts often have subsequent lines of security. Once a parasitic egg makes it into the nest, many hosts can reject it by throwing it out or flying off to dispose of the egg elsewhere. Rejection behaviour has two important components, the first being that hosts must be capable of telling the difference in colour and pattern between their eggs and the parasite's, and the second being to identify which eggs belong to whom (we'll come to that second component shortly). To beat this, the parasite should mimic the host's eggs. In some of his earliest work on cuckoos, Nick Davies teamed up with Mike Brooke in the 1980s and showed that many of the common cuckoo's host species do discriminate between eggs based on appearance, and that this has led to cuckoo mimicry.[8] Common cuckoos appear to show specific mimicry of the eggs of many of the host species that they utilize. They are able to do this because different individual cuckoo females specialize on just one species of host, such that the common cuckoo can be divided into a series of host races, often called 'gentes'.

Brooke and Davies analysed the appearance of cuckoo eggs from museum collections in the UK and scored the eggs for their level of spotting, brightness, and colour, determining that the eggs of the most common host races were indeed distinct in appearance. They then asked ten colleagues to match a selection of cuckoo eggs to those from four different hosts (reed warbler, meadow pipit, pied wagtail, and great reed warbler) based on their appearance. Generally, the subjects were able to do this accurately for many species. Finally, and perhaps most importantly, Brooke and Davies made model cuckoo eggs out of resin painted with different colours and patterns, many of which resembled the eggs of a different cuckoo host race, and placed them in the host nests (Figure 54). They then recorded which eggs the host parents rejected over a period of three days, and which eggs hosts accepted into their clutch. As predicted, most of the host species were more likely to accept eggs that resembled the cuckoo eggs of their own host race, showing that mimicry of cuckoos towards their targeted species does provide an advantage in tricking hosts. Dunnocks were

FIG. 54. Painted model eggs, from pioneering work by Davies and Brooke showing that many, but not all, hosts rejected non-mimetic eggs. This was just one of many experiments investigating brood parasites, testing if and when host birds reject foreign eggs. Left: the left-hand column are eggs of four hosts of the common cuckoo: dunnock, reed warbler, meadow pipit, and great reed warbler, with those in the middle column the corresponding eggs of the cuckoo host race. The eggs in the final column are the artificial eggs made by Brooke and Davies. Right: an artificial egg resembling those of a reed warbler (the larger one) in a reed warbler nest.

Left image Nick Davies and Mike Brooke; right image Nick Davies

the major exception because they accepted practically any egg, even ones of a very unnatural appearance (such as black eggs). Davies and Brooke ultimately tested twenty-four species of bird with model eggs, showing that when cuckoos have seemingly better egg mimicry host species tended to show stronger rejection behaviour. They also found that unsuitable hosts, such as those that feed cuckoo chicks an inappropriate diet and hence are not parasitized, don't reject eggs. Moreover, populations of birds in Iceland which rarely encounter cuckoos, such as meadow pipits and white wagtails, showed less rejection behaviour than populations of the same species in Britain where they are targeted by cuckoos.

Davies and Brooke's experiments pioneered the experimental study of egg mimicry and rejection behaviour and provided key evidence confirming how host defences select for mimicry in cuckoo eggs. However, their studies did have some drawbacks, most notably that human vision was used to score egg appearance and mimicry, and to create the model eggs. Birds, as we now know, have a somewhat different visual system from humans, including most likely a superior ability to see colours. Since Davies and Brooke's original work, many advances have been made in our ability to understand animal coloration from the perception of other animals. This includes various models of animal vision that scientists now employ to predict how other species see the world. In the last few years, several studies have used such methods to show that cuckoo eggs are often a close match to those of their hosts when considering avian vision, and that rejection behaviour by hosts is indeed driven by levels of colour difference between host and foreign eggs.

The Zoology Department of Cambridge University is famous for studies of cuckoos by Davies, Brooke, and many others. I was lucky enough to spend a number of years there too, including spending some time working on cuckoos. Mary (Cassie) Stoddard (a doctoral student at the time) and I decided to explore the question of just how well common cuckoo eggs really do match those of their hosts, through bird eyes.[9] To do so, we measured different bird eggs from the Natural History Museum in Tring in the UK. With several million eggs, this museum has arguably the largest collection of bird eggs in the world, including a vast collection of clutches of bird eggs that were naturally parasitized by cuckoos before being collected (Figure 55). By measuring the coloration of these eggs we could predict how the receptor cells used for colour vision in birds' eyes would respond to different egg colours, much like the model of bee vision we encountered in the orchid mantis study (Chapter 2). By so doing, we could compare mimicry of host eggs by different cuckoo host races in terms of a bird's visual system. Cassie and I found that cuckoo host mimicry was

FIG. 55. Mimicry by host races of the common cuckoo. Eggs along the bottom row belong to some of the most common host species used (left to right: great reed warbler, red-backed shrike, reed warbler, dunnock, brambling, garden warbler, meadow pipit, pied wagtail). Eggs on the top row are those of the corresponding host race of cuckoo. Generally the mimicry is good to excellent, but with some notable exceptions, such as the dunnock.

Images taken by Mary Stoddard and copyright The Trustees of the Natural
History Museum, London

often exceptionally good, including in the UV range, which birds can see but we cannot. Moreover, host species that are more likely to reject foreign eggs correspondingly have cuckoo host races with more sophisticated colour mimicry to try to defeat their discriminating hosts. We also found very similar results when we used digital image analyses to characterize the actual patterns and markings on the eggs: hosts with stronger egg-rejection behaviour had cuckoos with better pattern mimicry. For example, bramblings show extremely strong rejection behaviour and consequently have a cuckoo host race with an almost perfect match for egg pattern. In contrast, meadow pipits are not especially determined rejecters and their cuckoos are not close mimics for pattern. This tells us that there is not always a need to be a perfect mimic, but instead the degree of deception is often strongly determined by how much selection pressure is imposed by the species being duped. Presumably at some point, when hosts become highly discriminating cuckoos should benefit from changing to exploit a new, more naive host. How and when this happens is not clear because we do not know much about the mechanisms by which female cuckoo chicks that have been reared by a host select the same host species to parasitize later in life, and how fixed any mechanisms are.

This work, and other recent studies, supported Davies and Brooke's seminal experiments while at the same time accounting for how birds might see the eggs. However, what was still missing was a study that experimentally picked apart exactly how host parents discriminated between the cuckoo eggs and their own, and the specific aspects of egg appearance that they used to do this. Fortunately, as I mentioned already, the University of Cambridge is full of people who study brood parasites, including Claire Spottiswoode. Claire is an expert on African ornithology and a remarkable field worker. From South Africa, she has an in-depth knowledge of the birds and how to get things done in parts of Africa where working can be difficult. She set up a field site in Zambia where multiple species of brood parasite are found. One of these is the African cuckoo finch (*Anomalospiza imberbis*), a yellow-brown bird sometimes seen in flocks. Claire had been studying the cuckoo finch and some of its main hosts and was intrigued by the fact that many species had extremely diverse egg colours and patterns. In the most common host, the tawny-flanked prinia (*Prinia subflava*), females each lay one egg type from a range of stunning coloured eggs laid by the species as a whole—either red, olive, blue, or white eggs overlaid with a plethora of spots and squiggles (Figure 56). Each individual prinia lays the same egg type throughout her life, but individuals can differ greatly in egg colours and patterns, resulting in high levels of diversity in the species. The host race of the cuckoo finch does exactly the same thing, with females each laying eggs of different appearance.

I've been fortunate enough to work with Claire on several studies of the cuckoo finch and its hosts, including determining how hosts reject foreign eggs.[10] In the first study, Claire conducted experiments in Zambia that involved swapping an egg from the nest of one tawny-flanked prinia into the nest of another individual. Essentially, this involves playing the part of the parasite and simulating a cuckoo

FIG. 56. Mimicry in cuckoo finch eggs. Left: diversity in colour and pattern of eggs of the tawny-flanked prinia (*Prinia subflava*), outer circle, and those of its parasite, the cuckoo finch (*Anomalospiza imberbis*), inner circle. Individual females lay different egg colours from one another, leading to substantial levels of polymorphism in each species. Top right: a cuckoo finch chick (larger individual) with a prinia chick. The parasitic chick begs intensely, so that any host chicks normally die of starvation and only parasitic chicks fledge from the nest (bottom right).

Images Claire Spottiswoode

finch egg in the nest by using prinia eggs instead. Due to the high variation in prinia egg coloration among individuals, some matches to the host eggs were very good and some relatively poor. Over a period of days Claire checked the nests to determine which eggs were missing and therefore rejected and which were accepted (Figure 57). We then applied models of bird colour vision and image analyses of pattern to calculate on what basis hosts accepted or rejected eggs. The results were clear: prinia were more likely to reject eggs when the level of mimicry was worse, and they used several different aspects of egg appearance to determine whether an egg should be rejected. They based their rejection decisions on how different the host eggs were from the foreign ones for colour, and several aspects of pattern (including the size and diversity of the markings). Perhaps most

FIG. 57. Experiments revealing what features of egg colour and pattern host birds use to detect and reject foreign eggs. One approach has been to swap an egg from the nest of one individual into another and monitor if the host parents accept or reject it. The top image shows a clutch of prinia eggs containing two of the host's eggs and one egg from another female (right egg). In this case the mimicry was good and the foreign egg was accepted. The bottom image shows a clutch in which the foreign egg (right) was rejected.

Images Claire Spottiswoode

importantly, we found that the specific features hosts used in rejection corresponded precisely to those aspects of egg appearance that differed most between real cuckoo finch and prinia eggs. This means that the prinia are using the most reliable aspects of egg appearance, ones that convey the most information about identity (host or parasite), to guide their rejection behaviour. This likely has broad implications for the evolution of deception because it suggests that discrimination by hosts towards specific egg features leads to ever more effective mimicry regarding pattern and colour. Hosts become attuned to those aspects of egg appearance that are less well mimicked by parasites and use this to guide their rejection behaviour. In turn, parasites should improve their mimicry for these specific egg features. Therefore, mimicry and deception is not about a perfect match from the outset, but rather occurs in stages, with the deceived species (in this case the hosts) focussing on traits less well mimicked by cheats. This causes mimicry to be refined for some traits first, such as colour, rather than others like the shape of the egg blotches. If mimicry evolves in this way then it could also help to shed light on why imperfect mimics exist in other systems, such as Batesian mimicry by hoverflies.

I mentioned that successful rejection behaviour has two main components. As we've discussed, the first involves spotting the difference between host and parasite eggs. However, this is not enough because even when hosts have detected an egg that looks different, they still must identify which eggs are theirs and which are the parasite's. The host must reject the correct egg(s) rather than its own. Broadly, there are two main ways hosts could do this. On the one hand, they could simply throw out eggs in the clutch that are in the minority (an odd one out rule). If a bird has four eggs, three of which look very similar and one looks different, then the likelihood is that the disparate egg is the parasitic one and it should be removed. This approach requires little cognitive skill or learning because it's a simple rule of thumb, but it's a bad idea in circumstances when parasitism is so high that parasitic eggs could equal or outnumber host eggs (this does happen, with some hosts suffering several parasitic eggs in their nest). In such cases, the host could end up throwing out its own eggs instead. The alternative is that hosts 'know' what their own eggs look like, perhaps by learning their appearance during early breeding attempts, and then reject any eggs that deviate from this regardless of relative proportions in the nest. This method is less likely to result in errors because hosts should still reject the parasitic eggs even when they are more common than the host's. However, it requires more cognitive skill in order to compare the current eggs in the nest to some learned template of what the host thinks its eggs look like.

Generally speaking, it's the latter approach that seems to be the primary mechanism in most hosts. The first clear evidence came from the pioneering work of Stephen Rothstein in the 1970s on potential hosts of cowbirds in North America.[11] He placed model eggs of different numbers in the nests of hosts and showed that in most cases hosts could recognize their own eggs and reject the foreign ones, even when their own eggs were outnumbered. Similar findings have been found in hosts of a number of other brood parasites, including the prinia,[12] and while some studies have found evidence of birds using an odd one out rule, this seems to be far less common.

Egg rejection is far from the end of the story. Parasites have many other tricks to stop hosts from rejecting foreign eggs. This includes something rather nasty, whereby a parasite could bully hosts into accepting its eggs. The brown-headed cowbird (*Molothrus ater*) is the only widespread brood parasite in North America and has expanded in range considerably in the past two centuries. It is a very generalist parasite, exploiting more than 200 species, some of which show rejection behaviour while others do not. Lack of rejection could simply occur if some new hosts have not had the time to evolve defences (this is thought to be the reason why dunnocks don't reject cuckoo eggs). Alternatively, non-rejection might actually be the best option if parasites punish hosts for rejecting their eggs, and there is evidence that brown-headed cowbirds use these so-called 'mafia' tactics.[13] Female cowbirds continue to monitor host nests even after laying their eggs, and apparently retaliate against hosts that reject them by destroying the nest or chicks. This causes a complete loss of the host brood, forcing the host to re-nest, a process often followed by being parasitized again. Thus hosts can be forced into accepting parasitic eggs in order to avoid nest destruction. This strategy only works for the cowbird because its chicks do not normally evict the host young from the nest. Instead, host chicks are sometimes reared and survive alongside them, especially in larger host species. One or two host chicks sometimes still fledge. If rates of parasitism are high enough, and cowbirds use mafia tactics, then hosts might actually do better by accepting parasitic eggs because that way they have a chance of at least raising some young rather than having their nest repeatedly destroyed. In fact, nests of rejecting parents may produce around 60 per cent fewer chicks than accepting parents. Some cuckoo species may also use mafia tactics, but cowbirds are especially interesting because they may also 'farm' their hosts. That is, cowbirds sometimes appear to monitor the state of potential nests. If they find a suitable breeding host that's too far along in the reproduction process (e.g. already with chicks) then they may destroy the nest, forcing the hosts to re-nest, at which point the cowbird can target them. Interestingly, female cowbirds have an enlarged area of the brain called the hippocampus, which is widely known to play a role in spatial memory in birds and other vertebrates. This enables them to monitor the state of nests in an area and their outcomes by revisiting them periodically.

Until recently, evidence for co-evolution and deception in parasite–host systems beyond the egg stage (instead occurring with chicks or nestlings) was weak. Little evidence existed that hosts ever reject chicks, and there was correspondingly little indication of chick mimicry by parasites. Intriguingly, however, around the start of the new millennium evidence began to accumulate that some hosts do in fact throw cuckoo chicks out, especially in species that until that point had been somewhat understudied. Naomi Langmore, now at the National University of Australia, and colleagues at Cambridge were studying cuckoos found in Australia and started to make a series of interesting findings.[14] They investigated a host species called the superb fairywren (*Malurus cyaneus*), a small bird in which males have beautiful blue and black plumage, which is targeted by the Horsfield's bronze-cuckoo (*Chalcites basalis*). Langmore and the team found that fairywrens abandon almost half of all nests that contain a single cuckoo chick. Since the cuckoo evicts all host offspring, this behaviour could be explained through a simple rule of thumb: abandon nests with a lone chick because there's a high risk it's a cuckoo. However, this cannot be the full story because fairywrens abandoned 100 per cent of nests with a lone shining bronze-cuckoo chick (*Chalcites lucidus*), which is an infrequent parasite of the wrens. This led Langmore and colleagues to suggest that hosts might discriminate between parasitic young because hosts were less likely to desert nests with a lone Horsfield's bronze-cuckoo chick (or a lone fairywren chick) than a shining bronze-cuckoo chick, rather than simply deserting based only on chick numbers. Fairywrens may be less efficient at abandoning nests with the common Horsfield's bronze-cuckoo chicks if those cuckoos somehow resemble the host chicks more than the shining bronze-cuckoo chicks.

This was tantalizing evidence that chick rejection might exist, but it was not until a study by Japanese scientists led by Nozomu Sato from Rikkyo University in 2010 that clear evidence arose, involving the remarkable documentation of host parents throwing cuckoo chicks from the nest.[15] Sato and the team also studied Australian birds, and filmed large-billed gerygones (*Gerygone magnirostris*), which are parasitized by the little bronze-cuckoo (*Chalcites minutillus*), rejecting cuckoo chicks. Subsequent work found this behaviour also occurring in the mangrove gerygone (*G. laevigaster*), another host of the little bronze-cuckoo. So far, not much else is known about this behaviour, other than that it seems relatively infrequent and unrefined, with hosts often mistakenly rejecting their own chicks. Nonetheless, Langmore and her colleagues have shown that in several Australian species (including superb fairywrens), rejection of cuckoo chicks seems sufficiently strong a factor to have led to the evolution of remarkable mimicry[16] (Figure 58). Here, cuckoo chicks show a close resemblance (in terms of both bird and human vision) to the skin colour, gape colour, and presence of downy feathers of host chicks, with different cuckoo species resembling their respective hosts. For example, chicks of the little bronze-cuckoo match the dark skin of large-billed gerygone, shining bronze-cuckoo chicks are remarkably similar in yellow colour to those of the yellow-rumped thornbill (*Acanthiza chrysorrhoa*), and Horsfield's bronze-cuckoos are a pink flesh colour, just like superb fairywren chicks. This mimicry would only evolve if parents

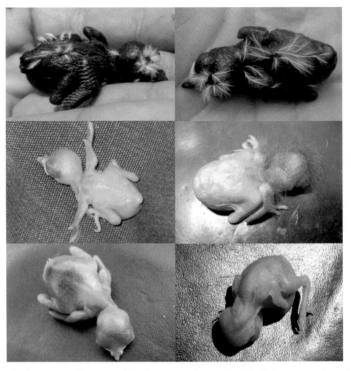

FIG. 58. Chick mimicry by Australian bronze-cuckoos. Top: little bronze-cuckoo (*Chalcites minutillus*) and its host the large-billed gerygone (*Gerygone magnirostris*). Middle: shining bronze-cuckoo (*Chalcites lucidus*) and yellow-rumped thornbill (*Acanthiza chrysorrhoa*). Bottom: Horsfield's bronze-cuckoo (*Chalcites basalis*) and superb fairywren (*Malurus cyaneus*). The cuckoo chicks mimic both the skin colour and the presence of downy feathers.

Images Naomi Langmore

discriminate between the appearance of different chicks. At the moment not enough is known about other parasitic systems to say whether there's something unique about these Australian birds and whether chick rejection might be more widespread in other groups too.

There's another puzzling aspect to many of the Australian parasite–host systems: hosts rarely seem to reject eggs. This is odd because why should hosts wait until the chick stage to reject given that most other parasites we know of focus on rejecting eggs? In addition, waiting until the chick stage invariably means the entire host's young will die when the cuckoo chick evicts them. In some cases, a lack of egg rejection by hosts occurs in species with very dark nests, making it almost impossible for hosts to see the difference between their own eggs and those of a cuckoo. But more recent work suggests that hosts do begin their defences at the egg stage, but they do not act on them until the chicks hatch.[17] Diane Colombelli-Négrel, from Flinders University in Australia, and colleagues also studied superb fairywrens and Horsfield's bronze-cuckoos. They found that the parent fairywrens sing a special song to chicks while they are still inside the egg, and that the unborn chicks learn this 'password'. The password differs among host nests, and host chicks incorporate it into their begging call when they hatch. The cuckoo, however, doesn't lay its egg until only around two days before the chicks hatch, by which point the fairywren mothers have finished teaching the special song. When the cuckoo hatches, it fails to sing the right tune and the parents feed it less, and even abandon the nest, leaving the cuckoo to die. This remarkable discovery shows that parental defences can span several stages of the reproductive process. What's less clear is if and how such passwords link with chick rejection and mimicry. It may be the case that fairywrens abandon lone chicks on the basis of a mismatched cuckoo begging call, whereas in other cases hosts directly analyse chick appearance and use that in physically rejecting chicks from the nest. Why they sometimes do one or the other, and how these two approaches link, remains a mystery, as does why these Australian birds differ so much from other parasites that have been studied.

Let's now move on to what Dawkins and Krebs called exploitation of hosts, that is, the way that parasites maximize parental care. In breeding birds there is often a conflict of interest between parents and chicks, with parents wanting to limit their provisioning to a certain level in order to 'save' some reproductive effort for potential future broods. However, chicks often try to manipulate parents into bringing more food than the parents otherwise would, through elaborate begging displays. In evolutionary biology this is referred to as parent–offspring conflict. As we discussed, in most cases there is a limit to how selfish chicks should be because they share many of their genes with their current and potential future siblings. This, however, does not apply to brood parasites because they have very low relatedness with the host parents and their young (coming from a different species entirely), and so they should be extremely selfish and extract as much care as possible. And, as we'll find out, they have many tricks up their sleeve to do this.

To monopolize parental care and gain as much food as possible, a common and obvious route is to eliminate the competition in the nest: i.e. kill all the host chicks. Common cuckoo young are famous for this. The chick normally hatches before the host's young and then, while still naked and blind, proceeds to heave the host eggs one by one on to its back and throw them from the nest on to the ground or into the water below. The host parent, seemingly bemused by the whole process, often just sits by and watches as all its eggs are pushed from the nest. This might sound nasty, but other parasites do things in a more gruesome manner. Honeyguides are most commonly found in Africa and Asia, and one species, the greater honeyguide (Indicator indicator) from tropical Africa, is famous for its mutualistic relationship with humans. The birds produce a special call when they find a bee nest, alerting and guiding humans to it in order to break it open and extract the honey. If the humans fulfil their share of the bargain then they leave a piece of the nest for the bird, which wants not the honey but rather the wax and grubs. But honeyguides, including the greater honeyguide, are also brood parasites, targeting species that nest in holes and cavities. Claire Spotti-swoode has studied honeyguides in Zambia and with Jeroen Koorevaar documented the way that the chicks eliminate the competition.[18] She put infrared cameras inside the nests of several host species that contained a honeyguide egg, such as the little bee-eater (Merops pusillus), which often nests in disused aardvark burrows, and filmed what happened next. Honeyguide chicks always hatched first, and as the host chicks hatched they attacked them vigorously using a sharp, specially adapted bill hook (Figure 59). The parasitic chicks proceeded to hack, stab, bite,

FIG. 59. Chick killing by greater honeyguide young. The top image is a honeyguide chick, aged about eight days, showing the bill hooks used to kill host chicks. The bottom image shows the honeyguide chick with three recently killed little bee-eater hatchlings, inside the underground nest chamber.

Images Claire Spottiswoode

and shake the host young until they were fatally injured, taking between around ten minutes and seven hours to die from internal bleeding. Claire noted that the honeyguide chicks would even attack and bite human hands that held them (but funnily enough, not the host parents feeding them).

It does not always pay for parasites to eliminate all of the host's chicks though. Surprisingly, they sometimes do better when reared alongside them. Brown-headed cowbirds, for example, don't evict host young but instead compete with them in the nest. Rebecca Kilner (another Cambridge brood parasite expert) and colleagues have shown that this is actually beneficial to cowbird chicks because hosts usually modulate their investment and bring more food to a nest with several chicks present than to a lone chick.[19] The cowbird still receives most of this food by begging more vigorously, so the cost of competing with the host chicks is small. Nonetheless, many parasites, including common cuckoos, do evict all the host young, and this could result in less food being brought by the parents. As with the Japanese hawk-cuckoo, the chicks of many parasitic species have a number of tricks to make up for this shortfall.

Let's again return to common cuckoo chicks, Nick Davies, and Wicken Fen. Davies and colleagues, including Rebecca Kilner, tested what caused reed warblers to bring more or less food to cuckoos.[20] Specifically, they built upon suggestions that the begging displays of cuckoo chicks seem much more intense than those of a single host chick alone—in fact, more like a whole brood of chicks. Their experiments tested how much food reed warblers brought to the nest when it contained a single reed warbler chick, a single blackbird or thrush chick (of a similar size to cuckoo chicks), or a single cuckoo chick. Warblers brought more food to the cuckoo chick than to the other chick species, suggesting that size alone was not key, but that instead something about the cuckoo begging display must matter. Analysis of the cuckoo begging calls supported this, showing that a single cuckoo chick vocalizes at a similar rate and intensity to a whole brood of reed warblers. Next, the team placed a single blackbird chick in a warbler nest and played recorded begging calls of either a cuckoo or an entire reed warbler brood through a speaker next to the nest (Figure 60). This time there was no difference in the amount of food hosts brought to the nests, showing that common cuckoos seem to maximize provisioning by sounding like an entire brood of host chicks.

FIG. 60. Playback experiment, used to test how host parents respond to the begging calls of their own young and cuckoo chicks. Here a speaker is placed next to a reed warbler nest containing a single blackbird chick. Parents have high rates of food provisioning when the speaker plays the begging calls of a cuckoo chick.

Image Nick Davies

The begging calls of the common cuckoo have similar characteristics to calls of a whole reed warbler brood, and for that reason have often been considered as an example of acoustic mimicry of an entire brood. However, this may not actually be correct.[21] A point we have returned to several times is that mimicry involves one object or stimulus (a cuckoo call) evolving to resemble another (a warbler call) closely enough that a third party (a warbler parent) mistakenly classifies the mimic as another object type or species (just as hoverfly colour patterns evolve under selection to be mistaken for a wasp or bee). The problem with considering cuckoo begging calls as mimicry is that hosts do not seem to use begging calls to classify the cuckoo as belonging to one object/species or another. That is, there's no evidence that reed warblers use differences in begging calls to discriminate between host and cuckoo chicks and use this information in rejection. It's more likely that the cuckoo begging call is another case of sensory exploitation instead, that is, simply a way of getting more food. The function of the begging display is not to prevent detection and rejection by the host parents, but is instead to extract as much care as possible from the parents by causing a heightened feeding response. In effect the cuckoo evolves a highly elaborated 'supernormal' version of the reed warbler call because this should be more effective in stimulating the host parents into bringing as much food as possible. The reed warbler begging call has already evolved to do this; it's a successful call that the parents respond to effectively, and so the cuckoo call converges on an exaggerated version of this. As a result, superficially it looks like mimicry, but it's more a case of evolutionary convergence, with the exception that the cuckoo calls more intensely because it's more selfish and needs more food.

In contrast to most other begging displays, one likely instance of genuine mimicry to increase host provisioning occurs in the Japanese Horsfield's hawk-cuckoo, the species from Mount Fuji that we encountered at the start of this chapter. The bright yellow wing patches are highly visible, but Keita Tanaka and others also found that hosts sometimes try and directly put food into the wing patches, illustrating that they misclassify them as real chick mouths.[1] The mimicry to human eyes is far from perfect, but under dark nesting conditions it may be more than sufficient. In fact, the wing patches seem to use both sensory exploitation and mimicry because to avian vision they are brighter than real host mouth colours. Furthermore, the wing patches reflect high levels of UV light, which matters because the high-altitude forest environment in which the birds live is likely to contain lots of UV light (Figure 61).[22] While the common cuckoo uses an exaggerated

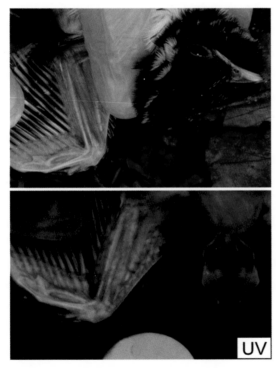

FIG. 61. The wing patches of the Horsfield's hawk-cuckoo young. Chicks have bright yellow wing patches, which are especially bright in reflecting UV light.

Images Martin Stevens

vocal display to bring food, the Japanese hawk-cuckoo uses a visual signal instead, probably because their hosts are ground nesters and suffer from very high levels of predation (over 50 per cent fail), and calling loudly would increase this risk.

Birds are not the only group of animals that cheat on one another for reproduction. Ants, bees, and wasps (the Hymenoptera) also do this widely and enlist a range of trickery and deceit similar to birds in doing so.[23] In contrast, however, deception in insects primarily occurs in the chemical rather than the visual sense, which reflects the importance of this mode of signalling to many of these species. Brood parasitism, more often called social parasitism in insects because it frequently involves exploiting species living in colonies, generally takes two forms. In ants, more than 200 of the approximately 12,000 species are known to be social parasites. One type of these exists as so-called inquilines, which invade the nests of their hosts and live alongside them, sometimes killing the host's young or manipulating the host workers into raising the parasite's offspring instead. We already encountered another type of inquiline at the start of the book with the blue butterfly caterpillars. The other group of social parasite is the slave-maker ants that capture the young of other species and force them to do their bidding.

Perhaps the most remarkable social parasitic ants are the slave-makers. As the name implies, these attack and steal the brood of other colonies, usually of a different species, bringing them back to their own nest and tricking them into working for the host. Darwin suggested in the *Origin of Species* that slave-maker ants evolved from species that were predators of the eggs and larvae of other ants. While it is not clear if this is correct, it's a sensible prediction, and slave-making seems to have evolved several times independently, demonstrating its advantage as a way of life. Slave-maker ants have many specializations associated with their life history, some of them very extreme. In the first instance, a new colony often starts when a queen invades another nest. For example, the European slave-maker ant (*Harpagoxenus sublaevis*) targets two small species: *Temnothorax acervorum* and *T. muscorum*. Parasitic queens mate and then invade a colony of one of their hosts, before killing or expelling the current queen and all adult workers. The developing host brood imprint on the new queen's odour and when they hatch perform the colony tasks, including rearing the parasite's own young. These then replenish the host worker force by conducting slave raids, capturing broods from other host colonies. Sometimes slave-maker nests raid and even contain workers from both *Temnothorax* host species. Analysis of the surface chemical profiles on the bodies of individuals that enable recognition (so-called hydrocarbon profiles) shows that while the adults have very different chemical profiles, the pupae are often quite similar and therefore seem to acquire the odour of their colony as they mature.[24] This explains how it's possible to integrate broods from different species and colonies into the same nest. Some slave-maker species apparently steal broods from several species in order that each undertakes different tasks in the nest, just like a division of labour that often occurs in ant colonies with different castes.

Slave-maker workers are often built for one thing alone: to attack and raid the nests of other ants and steal their brood. A single colony of slave-makers can capture thousands of broods in a season. They are remarkable fighting machines, built with an arsenal of weaponry. In the first instance, scouts from the slave-maker colony are sent out to look for nests of their host species. When found, the scouts return and recruit the army to attack and steal the larvae and pupae of the hosts. Taking it to extremes, sometimes slave-makers are even accompanied by enslaved workers as well, who assist in re-raiding their original homes. On arrival in the area the slave-makers search for the target nest, and when it has been found they remove any obstacles such as small stones blocking the entrance. Then they enter the nest. When this happens, workers of some species flee, whereas others, especially in larger well-defended colonies, stay and fight. The latter often results in considerable casualties due to the weaponry the attacking ants possess. For instance, many slave-maker workers are equipped with powerful mandibles, used for snipping off the heads of workers from other species. Next, they use chemical warfare, which takes a wide range of forms. In some cases slave-maker workers mimic their hosts' odour profiles to sneak past the defences. Frequently though they use a specially enlarged organ called the Dufour's gland to spray substances of varying effect at the hosts. Some of these are 'propaganda' substances, first described in slave-maker ants in the early 1970s,[25] which serve to appease the hosts, nullifying their attack behaviour. Other chemicals simulate host alarm signals, causing hosts to panic and flee. But the most remarkable ones are chemicals that actually cause the host workers to turn on and attack each other.[26]

Just like their avian counterparts, many duped species fight back against social parasites. Ants of the same colony often share the same chemical signature, on to which young imprint when they hatch. This enables recognition of nest mates and detection of intruders and competitors. Some slave-maker ants have chemical profiles that resemble those of their hosts, and this match is often closer to their main host species and to colonies occurring in the same geographic region than to host colonies from elsewhere (just like the blue butterflies).[27] This chemical mimicry enables queens to take over a nest by avoiding aggression, and for workers to evade host defences during slave raids. But in much the same way that cuckoo hosts can become more discriminating towards foreign eggs, ant colonies that are under heightened risk of attack are more discriminating towards chemical signatures that deviate from their own.

Alexandra Achenbach and Susanne Foitzik from Ludwig Maximilian University, Munich showed how hosts use chemical discrimination to fight back.[28] They found that *Temnothorax* worker ants enslaved by the slave-maker species *Protomognathus americanus* sometimes neglect the slave-maker brood, leaving two-thirds of them to die, or even killing them directly. Initially, this might not seem especially helpful because the colony has already been taken over and the host queen killed. However, quite often neighbouring nests of the same species have individuals that are relatively closely related owing to the way that new colonies form. By killing the slave-makers' brood, hosts can reduce the effectiveness of their parasites at raiding other neighbouring ant nests, and in the process benefit individuals to which they are related. The help can be substantial because *P. americanus* can successfully raid between two and ten colonies per year, sometimes more, placing substantial pressure on the host population. Through hosts killing parasitic larvae the number of slave-maker workers is kept low. In these species it actually seems that the chemical profiles of host and parasitic larvae are still quite different, and this allows enslaved workers to discriminate between them. However, the researchers also found that when they swapped parasitic pupae into host nests, pupae were more likely to survive when placed in host nests from the same area than when put into nests from elsewhere. This shows that although the chemical profiles of the parasites are not a perfect match to those of their hosts, the slave-makers have undergone some local adaptation and their chemical profiles are more similar to those of local hosts than to those from elsewhere.

Another line of defence is to physically stop raids from occurring, including by placing guards on nest entrances and attacking slave-maker scouts looking for nests to raid. Other work by Foitzik and colleagues on *P. americanus* and its host *Temnothorax longispinosus* explored how this depends on the risk of attack.[29] They collected colonies from different regions of North America and induced slave raids between colonies from different locations. Host ants from locations where parasitism was more common showed greater defensive behaviour, including heightened aggression towards parasitic scouts, than ants from places where slave-makers were rare. Correspondingly, parasites from regions where slave-making is common were better at attacking host colonies. Those ants used strategies such as placing an ant close to the host nest entrance to keep it clear, facilitating the raiding ants to escape. In locations where slave-making is common, hosts have heightened defences, forcing parasites to have enhanced attack strategies.

Inquiline social parasites have many traits in common with slave-maker ants, including how hosts fight back. For instance, a tiny ant called *Temnothorax kutteri* from Germany and Sweden also uses propaganda substances when the queen invades host colonies. An invading queen daubs the host workers with a clear substance from her abdomen, again produced from the Dufour's gland.[26] This makes the host workers (*T. acervorum*) turn on and attack each other, allowing the queen to enter the nest. Once inside, the queen may gradually acquire the host odour in order to be accepted by the workers. This process of odour acquisition is probably a common approach, a bit like starting off with a blank slate lacking any significant hydrocarbons (a tactic called 'chemical insignificance'), followed by grooming with host workers or touching parts of the nest, smearing themselves with the host colony odour. Other species of inquiline ant have different tactics to gain initial entry to the nest. For example, some queens (e.g. *Anergates atratulus*) apparently play dead, tricking the host workers into carrying them inside.

It's not just ants that are social parasites—bees and wasps also get in on the act. Bates and Wallace were very much aware of the 'cuckoo bees', with Wallace noting that they often resemble the appearance of species that they target.[6] In fact, 15 per cent of bee species worldwide may actually be parasitic, with the strategy likely arising many times during evolution. For example, female *Sphecodes* cuckoo bees lay their eggs in the nests of other solitary bee species. I've been lucky enough to have some exploiting bee nests in my garden at the same time as writing this book, and to observe their behaviour as they do so (Figure 62). Like

FIG. 62. Many bees are 'cuckoo' species, laying their eggs in the nests of other species and exploiting their resources. Here, the solitary bee *Lasioglossum calceatum* (top) is being used by the cuckoo bee *Sphecodes monilicornis* (bottom). The cuckoo visits the burrows of the host while they are away.

Images Martin Stevens

many avian cuckoos the bees are often secretive, waiting in the vicinity of a host nest and only entering when the host is away. If they are successful then the cuckoo bee larvae hatch and destroy the host's young with pincer-like mandibles (if the adult bee has not already done this during laying), allowing the parasites to feed off the pollen supplies in the nest, just as avian cuckoos kill the host chicks before monopolizing parental care. In another parallel with avian parasites, some cuckoo bee species from Europe exploit over ten host species (as opposed to some parasites that specialize in just one or two hosts).[30] However, each female seems to specialize in just one host, akin to host races or gentes in birds. Some thirty species (of over 250 in total) of *Bombus* bumblebee are also social parasites.[31] Invading queens carry a number of defences in case they get caught, including powerful stings and large venom sacs, sharp mandibles, as well as a thick body cuticle to protect them from host attacks. Some species also have an enlarged Dufour's gland to secrete chemicals that repel the host workers. Some *Bombus* cuckoo bees also mimic the chemical profiles of their hosts to circumvent defences, with hosts attacking invaders more violently when the mismatch is greater.

Brood and social parasites are relatively widespread among birds, bees, wasps, and ants. But they are seldom known elsewhere. Remarkably, some fungi exist in the form of 'termite balls' that mimic the shape, size, and even odour of termite eggs in order to be cared for by the workers ('fungal cuckoos').[32] A bit like the ants rearing blue butterflies in Chapter 1, the termites bring the fungal balls back to the nest, where they germinate. The relationship is probably parasitic and certainly deceptive by the fungus because the termites waste significant time looking after the fungal balls, whereas the fungus is protected from drying out, and from competitors and pathogens. However, beyond a sporadic collection of known examples such as this, why is social and brood parasitism largely restricted to so few select groups of animals? The answer is not really known, but one thing the insects and birds have in common is the rearing of eggs in a nest. Mammals (echidnas and platypus aside), for example, have internal gestation so brood parasitism is not an option. There is also no potential for brood parasitism to occur in the many vertebrate and invertebrate species that simply lay their eggs and leave them because there is no parental care to exploit. But other animals do make nests, lay eggs, and care for them, including various fish, frog, and reptile species.

At least one brood parasitic fish does exist, a catfish (*Synodontis multipunctatus*) from Lake Tanganyika in Africa, which utilizes mouth-brooding cichlids as hosts, as reported back in 1986 by Tetsu Sato of Kyoto University in Japan.[33] Mouth-brooding cichlids pick up their eggs and small fry and retain them in their mouth for protection. The developing fry periodically leave the mouth to forage as they gain maturity, but return for protection. While studying the parental care of cichlids in Tanganyika in 1985, Sato noticed that a number of the female cichlids, from six different species he sampled, had catfish young in their mouths. The catfish, it seems, lays its eggs in the vicinity of its cichlid hosts, so that the hosts pick up the foreign eggs along with their own. As the young fish develop, the catfish young attack and devour the host fry while still in the mouth of the parent. Like an avian brood parasite, the catfish not only exploits the parental care of the host, but also kills the host's own offspring.

Much more recently, in 2010, another species of catfish from Lake Malawi was suggested to be a brood parasite.[34] *Bagrus meridionalis* is a catfish species in which both males and females care for the young: the female feeds them with unfertilized eggs and the male brings back small prey items as food. Their nests sometimes contain high numbers of young from another catfish, *Bathyclarias nyasensis*, which are fed and protected by the host parents as if they were their own. In fact, those nests contained almost exclusively individuals of the (potentially) parasitic species, perhaps indicating that the foreign young devour the host's own when they hatch. There are a handful of other possible instances of brood parasitism in fish, but generally it's unclear if this is costly to the hosts or whether it's simply a case of one species dumping its eggs in the nest of another with little negative cost to host reproduction.

Beyond this, why brood parasitism is not more widespread is hard to know. It may be more common than we realize. Cuckoo eggs in a host nest, and enslaved workers in ant nests, are relatively easy to see through their different appearance, and hence to be discovered by natural historians. But the eggs of most animals are not especially characteristic to each species and so brood parasitism would be easy to miss. There may be many instances of individuals putting their eggs in the nests of other individuals from the same species (indeed, such within-species parasitism is widely known in birds), and this behaviour could form a precursor to parasitizing other species. Detecting parasitism, and even knowing where to look, remains a problem, but the key point seems to be that it either occurs when the host species show considerable investment in parental care or when the parasite would have to do so were it to rear its own young. What's clear is that brood and social parasitism have provided a wealth of examples and theory to help us understand how deception works and evolves, and indeed how individuals of exploited species fight back. For one thing they illustrate how deception often occurs along a series of steps, often becoming more sophisticated in the process (for example the degree of egg mimicry). These stages often coincide with increases in host defence as the arms race escalates. They also show how animals such as hosts use and respond to available information in guiding their behaviours, for example in egg-rejection behaviour, and how parasites manipulate host responses by targeting their deception accordingly. The subject in general has also aided our understanding of how deception evolves to reflect the costs and benefits of different potential strategies, because parasitism is rarely cost-free once hosts start to fight back. However, what we have yet to figure out is why interactions between host and parasite can be so variable among species, and why some groups progress down one evolutionary pathway (such as egg rejection and mimicry) and others down another (for example chick rejection and mimicry).

8

SPREADING GENES AND
SEXUAL MIMICRY

. . . .

Throughout the book we have focussed mostly on deception geared towards improved survival, either through avoiding predation or obtaining food. However, in Chapter 7 we began to look at how animals use deception in reproduction, specifically by focussing on brood and social parasites, which exploit other species to rear their young. What matters ultimately in evolution is passing on your genes through reproduction, and there are many other types of deception beyond brood parasitism for achieving this. This includes tactics that exploit individuals of other species to help in the reproductive process, and it can also involve individuals of the same species tricking one another to gain an advantage in mating.

For organisms that reproduce sexually, male gametes (sex cells, e.g. sperm) must fertilize female eggs. This normally requires the gametes of one individual to be transferred to another, and in nature this can occur in a variety of ways. For many plants to reproduce, pollination, in which pollen is transferred from the male parts to the female parts in order for fertilization to occur, is a key process. Frequently, pollen is simply released into the environment, a sort of a scatter-gun approach, whereby enough of it is released such that some should land on the relevant parts of other plants of the same species in the environment. However, this process is rather crude and much pollen is wasted. Many flowering plants on the other hand utilize pollinator species to transfer pollen directly from one individual to another. The pollinator often gains a reward from the plant in the form of sugary nectar or the pollen itself. While we often think of pollinators as being mostly insects, other animals are also often involved, including birds and mammals (e.g. bats). It is insects, however, which often form the closest relationship with their targeted plants, even sometimes including some form of co-evolution between them. For example, insect species can show biases or preferences for specific flower colours, such as blue, and the flowers they visit often match that colour. Over evolutionary time, insects' preferences become stronger and the flowers become bluer.

For the plant, feeding pollinators with nectar and pollen carries costs, in terms of the energy required to produce those rewards. There should be a benefit to plants that can attract pollinators while at the same time not rewarding them. The orchid *Dendrobium sinense* is a species endemic to the subtropical Chinese island of Hainan. It is pollinated by a hornet, *Vespa bicolour*, which the orchid tricks into pollinating it in a most unusual way.[1] As well as feeding on nectar and pollen, hornets are ferocious predators of other insects, including other social Hymenoptera. They often attack honeybees, bringing them back to the nest to feed to the developing hornet larvae. The flowers of *D. sinense* are white with a red and yellow centre, superficially not dissimilar to the appearance of bees. Curiously, the flowers are pounced upon by visiting hornets in a behaviour resembling the way they capture prey. This is often enough for pollen to attach to the hornets, or for pollen to be deposited. But this is not the most inter-esting aspect of the deception. *D. sinense* also seems to produce chem-ical compounds resembling those found in the alarm pheromones of Asian and European honeybees, which can be detected by the hornets.

Thus, the orchid seems to lure in pollinating hornets by misleading them with a combination of chemical and visual mimicry.

Flower species that are permanently unrewarding are found in at least thirty-two different families of plant. However, maintaining cheating is actually not that easy because over time pollinators that go unrewarded should stop visiting deceitful plants. To prevent this, many flowers, like the Hainan orchid, have come up with all sorts of deceptive measures that tap into preferences that pollinator species often have, occurring in a range of sensory modalities, but especially vision and smell.

Deception in unrewarding plants can occur in broadly one of two main processes. First, plants could exploit a general preference that many pollinators have for bright, large, or colourful structures like flowers that stand out from a general green foliage background, by the (now hopefully familiar) process of sensory exploitation. That is, plants could exploit general, non-specific preferences that pollinators may have. Alternatively, or even additionally, plants could utilize mimicry, by resembling the appearance of a specific rewarding model species: either another plant, or something else entirely.

While permanently unrewarding flowers are probably relatively widespread, little is known about them outside one group: the orchids. Orchids are masters of deception, with many species duping pollinators in a remarkable variety of ways.[2] There are around 30,000 known species of orchid, of which approximately one-third are thought to use deception in some way for pollination. This involves mimicking cues that to pollinators would suggest the presence of food, potential egg-laying sites, or sexual partners. Overall in orchids, food deception by resembling some sort of rewarding flower that does provide nectar appears to be the most common tactic, though egg-site and sexual deception are also common. Sexual deception normally involves deceiving hymenopteran males (such as bees and wasps) into thinking that a female is around, inducing them to try to mate with the flower. This is perhaps the most widely studied deception orchids utilize. Many species have flowers visually resembling the general appearance of a female insect, inducing the male to try and mate with it and transferring pollen in the process. But chemical mimicry is perhaps more common, and some orchids mimic the sex pheromones released by female insects used to attract males. For example, the Australian species *Chiloglottis trapeziformis* is pollinated by a wasp, *Neozeleboria cryptoides*, and mimics a specific pheromone component of female wasps. Researchers have shown that treating black plastic beads with this orchid and wasp chemical signal induces male wasps to copulate with them.[3]

Sometimes mimicry by orchids can take elaborate pathways. *Epipactis veratrifolia* is a species found from the Middle East to the Himalayas, and south into Somalia and Ethiopia, and is pollinated by five species of hoverfly. The larvae of various hoverfly species attack and eat aphids. As such, one of the main factors determining where hoverflies lay their eggs is the presence of suitable food supplies. Because hoverfly larvae cannot disperse far, having plenty of aphids nearby is important. Female hoverflies can use chemical cues emitted from both the affected plants and the aphids themselves to detect not just the presence of aphids but even what species they are. The flowers of *E. veratrifolia* seem broadly to resemble the appearance of aphids, but again it is chemical mimicry that seems of primary importance in attracting hoverflies.[4] A project in Israel, led by Johannes Stökl at the Max-Planck Institute for Chemical Ecology in Germany, showed recently that the orchid produces chemicals that mimic the alarm pheromones of several aphid species produced when they are under attack from predators. Remarkably, this suggests that the orchid deceives hoverflies into visiting the plants and helping pollination under the false promise that there is a ready supply of aphids. Stökl's study used chemical analyses to compare the orchid emissions to those of aphid alarm signals, and found a close match between the two. They also made synthesized versions of the plant compounds and carefully recorded the neural responses made in the antennae of hoverflies in response to these chemicals, showing that the hoverflies should be able to detect these signals. They then showed that female aphids laid their eggs on bean plants in the lab when these were presented along with the orchid chemical compounds.

The deception by *E. veratrifolia* sounds very sophisticated, but just how specialist is it? The team suggest that it's actually quite generalist because the orchid emissions do not match the precise proportions of the different chemical components of aphid alarm signals. Instead, they seem to match the general presence of compounds found in the alarms of a whole group of aphid species. This actually makes sense because the hoverfly species the orchids attract don't seem to attack one species of aphid alone, but rather target several species. As Stökl and colleagues discuss, whether such generalized resemblance should be called mimicry or sensory exploitation is not entirely clear. On the one hand, we could argue that the situation is analogous to some theories of imperfect Batesian mimicry (such as one hoverfly species mimicking several species of bee or wasp), especially in cases where mimics evolve to resemble several models. Alternatively, the orchids might just be tapping into a generalist bias or perceptual preference that the hoverflies have for certain compounds because these reliably function as cues to the presence of food. Finally, quite how this type of deception evolved is also unclear, but one idea is that some plants (such as wild potato) contain chemicals similar to aphid alarm signals that may deter aphids from infesting them. *E. veratrifolia* might have taken this a step further by using the compounds to attract pollinators too, though this remains speculative. Sensory exploitation rather than specific mimicry also seems likely to occur in other orchid species. Research analysing the chemical compounds released by European orchids shows that some-times bees actually prefer to visit species with novel odours, rather than smells that accurately mimic bee pheromones more closely.[5]

Deception by orchids from afar to draw in pollinators is probably most often based on chemical cues. However, visual cues are likely to be important at close range, bringing the pollinator to the right part of the plant for them to be beneficial. Anne Gaskett from the University of Auckland and Marie Herberstein from Macquarie University in Austra-lia analysed the coloration of four species of orchid (from the genus *Cryptostylis*) that deceive male wasps of one species (*Lissopimpla excelsa*) by apparently mimicking the appearance of female wasps, with which the males try to mate.[6] To human eyes the orchid species look quite different and are not particularly convincing mimics of the wasps, bearing only a crude resemblance to their colour and shape (Figure 63). However, Gaskett and Herberstein used models of pollin-ator vision to calculate how the different orchid colours appear to the visual system of a wasp, which is generally poor at seeing colours of longer wavelengths (reds and yellows). Surprisingly, both the colours of the orchids and those of the wasp were essentially indistinguishable to the wasp's visual system, meaning that although the mimicry to us is poor, it's actually convincing if you are a male wasp looking for a mate. As mentioned previously, insects' visual abilities to see patterns and shapes is limited by their compound eyes, meaning that the wasps are probably not brilliant at telling the difference between the patterns on the orchids and the female wasps. Overall, the resemblance is probably good enough to fool the males, and because the male wasps are the only pollinators the orchids seem to use, they can specialize in mim-icking the colours that this species has, rather than generalist mimicry of multiple pollinator species. This example also tells us that, as has been suggested with Batesian mimicry, sometimes mimicry can be imperfect only to our perception, not to those species that matter.

FIG. 63. Deception by *Cryptostylis* orchids, which attract pollinating male wasps that try to mate with the flowers. Although the mimicry is not especially strong to human eyes, to the vision of a male wasp the colours of the flowers are indistinguishable from female wasps. Top left: *Cryptostylis leptochila*; top right: *C. subulata*; bottom left: *C. erecta*; bottom right: *C. hunteriana*.

Images Anne Gaskett

Orchids do not just deceive pollinators through sexual deception of animals, but also through mimicry of other plants. This can involve flowers mimicking the colours of other species that do produce nectar, and thus exploiting an honest communication system to the orchid's own end. Ethan Newman, from the University of Stellenbosch in South Africa, and colleagues studied a species of orchid from South Africa called *Disa ferruginea*, whose flowers attract pollinating butterflies despite offering no nectar reward.[7] The butterflies depend on gaining nectar from a few plant species. The authors showed that the orchid has orange flowers in the eastern part of its distribution but red flowers in its western range, and that the colour of these very closely resembles that of the corresponding model species. Furthermore, butterflies that pollinate the orchid prefer to visit artificial red flowers (mimicking real ones) and red orchids in the west, but prefer artificial orange flowers and orange orchids in the east. This corresponds with the preferred nectar plants of the butterfly being red in the west and orange in the east, and suggests that the orchid has evolved different flower colours to mimic these different model flower species, driven by local colour preferences of the butterfly they target. Thus, the selection pressure on the orchid to mimic another species of flower is sufficiently strong to lead to different colour forms of the same orchid species. Finally, as if mimicry of animals and plants were not enough, some orchids also mimic fungi. At least one orchid from south-western China (*Cypripedium fargesii*) appears to exploit pollinating flies with leaves that mimic fungus-infected foliage that the flies feed on.[8] The orchid flowers have blackish hairy looking spots that may lure the flies even though there is no real reward. Interestingly, the plant also produces a mild but unpleasant smell that has been described as resembling rotting vegetation.

Sexual deception for pollination is known almost exclusively from orchids, but it must surely occur in other plant groups too. Consistent with this, recent work by Allan Ellis from Stellenbosch University and colleagues have reported evidence that some South African daisy flowers from the species *Gorteria diffusa* use sexual deception to attract flies[9] (Figure 64). The flowers of this species are highly variable (polymorphic) in both shape and their yellow-orange colours, comprising something like fourteen different types from different geographic regions. Furthermore, the flowers have on the petals raised dark fly-like markings that even include bright patches resembling light

FIG. 64. The South African daisy (*Gorteria diffusa*) comes in a wide variety of forms from different geographic regions, some of which deceive pollinating flies. The raised markings on the inflorescences induce male flies to copulate with them, pollinating the flowers at the same time.

Images Allan Ellis

reflecting from the shiny wings and body, which seem to attract flies. In fact, flies are not only attracted to these markings but also attempt to copulate with them. The daisies do actually produce food rewards in the form of nectar and pollen, so are they truly detrimental to the flies? It seems that they are because the flowers are often very common, forming large carpet-like displays, meaning that the flies could spend considerable amounts of time visiting and trying to mate with flowers rather than real females. The resulting lost mating opportunities and wasted energy would be damaging to an individual fly's fitness. However, male flies are able to learn about the mimetic displays, becoming less likely to copulate with them through experience with the flowers.[10] Through this process they can reduce the costs to themselves, and the daisies may therefore rely more on naive flies for pollination. We might then expect the daisy to evolve better mimicry to continue to deceive flies for longer, followed by the evolution of flies with better abilities for learning and discrimination, resulting in a co-evolutionary arms race between them. Learning and resistance by the flies might also be the reason why the daisy is so polymorphic, because new varieties would initially be unknown to the flies and quickly spread in an area once they have arisen because initially the flies have not learned to avoid them.

Fungi are a large and highly varied group of organisms are fungi, but we have seldom mentioned them so far. This is largely because, despite their varied appearance, they have been less well studied in terms of communication with animals or specifically deception. However, it has been appreciated for a long time that many species of fungi attract animals to them with odour, including unpleasant-smelling species that mimic the compounds released from real decaying flesh, as is the case in the appropriately named 'stinkhorn' fungi (Phallaceae). The smell lures flies' towards a secreted slimy substance ('gleba') that the flies then eat. As they do so, they also ingest spores that later germinate after passing through the flies' digestive systems, or spores simply get stuck to their legs for transport elsewhere. The flies do not seem to gain substantial nutritional benefits, nor do they benefit in terms of places to lay their eggs (unlike in genuine rotting carcasses), and so they are largely exploited by the fungi. By using them, the fungus can help spread its spores to reproduce, rather than relying on the wind, as many species do.

Something in the region of seventy fungal species, from around 100,000 described overall, produce bioluminescent light. This has often been assumed to be a by-product of other processes, but a recent study by Anderson Oliveira from the Universidade de São Paulo, and colleagues from elsewhere in Brazil and the US, suggests this is not always the case.[11] They showed that the mushroom of a species of fungus (*Neonothopanus gardneri*) found in Brazilian coconut forests has an intense green bioluminescence produced only at night (Figure 65). Night time is when humidity is greater, aiding fungal spore germination, and also dark enough for the bioluminescent glow to be visible. This suggests that the light production is not just a by-product of other metabolic processes, because if that was true then it should occur continuously over twenty-four hours. When Oliveira and colleagues made fake glowing mushrooms from clear acrylic resin

FIG. 65. The fungus *Neonothopanus gardneri* in the top images glows with green bioluminescent light at night (right). Artificial fungi with green LED lights (bottom left) show that this glowing attracts insects that help to spread fungal spores.

Images Cassius Stevani

with LEDs, matching the light production of the real fungus but without the associated smells, a host of insects were attracted to them (but not to non-glowing control models), just like the real fungi. What we don't know yet is why insects are attracted to the glow, but presumably some sort of sensory exploitation or luring based on behavioural attraction to light at night is involved. Whether or not the fungi are truly deceptive is also unclear, in terms of whether the insects derive any reward from their visit or just waste time that could be devoted to other tasks.

While reproduction is key to success in evolutionary terms, the interests of individuals of the same species do not always coincide. For example, many female animals can produce only a limited number of egg cells, often spaced out in time. In contrast, males frequently produce copious amounts of sperm, far more than could ever be successful in fertilizing eggs. This means that females benefit not just from mating successfully, but also through choosing high-quality mates. For males, however, the premium is more on getting as many successful mates as possible because mating with a genetically 'poor' female does not really affect them from, in principle at least, getting many other matings with good females as well. In reality the exact situation varies a lot from species to species, and males can be choosy too. Nonetheless, we often end up with situations in which males go to great lengths to persuade females to mate with them, including through deception.

Darwin was the first to truly appreciate the importance of female choice, arguing in his seminal 1871 book, *The Descent of Man and Selection in Relation to Sex*,[12] that many elaborate mating displays had evolved through his new theory of sexual selection, driven by female choice. Darwin argued that females possessed some sort of aesthetic sense, preferring males with brighter, louder, and more elaborate displays. This idea—that females could choose their mates and be the drivers of remarkable displays such as the stunning colours and elaborate plumage of male birds of paradise or the peacock—was too much to bear for many in conservative Victorian society. In fact, it took almost ninety years before the idea started to become widely accepted by scientists, and before the first significant experimental evidence accumulated. Today, there is no doubt that females of many species, both vertebrates like birds and reptiles, and invertebrates like butterflies and spiders, do impose considerable choice on males, leading to the evolution of varied mating displays and male ornaments. Quite why females choose is still debated, but multiple, non-mutually exclusive ideas exist that broadly encompass two main benefits to them. On the one hand, females could gain direct rewards by mating with a high-quality male who is likely to be good at things such as provisioning and protecting their young. Alternatively, females could obtain indirect genetic benefits because good males should have good genes, leading to offspring with genes better suited to survival or gaining mates them-selves in the future. What Darwin and most other scientists for many decades did not know was the extent to which males can exploit and manipulate female preferences using a variety of tricks to gain their favour, through strongly stimulating or even directly deceiving female sensory and cognitive apparatus.

Bowerbirds provide an illuminating example of how males manipu-late female perceptions to gain their favour during mate choice. In many species the male builds a conspicuous three-dimensional structure, a bower, often made from branches and twigs on which he displays to females. In some birds, the male adds a wonderful variety of decorations, including vivid blue or green objects from the environment (sometimes including discarded human objects). Females visit the bowers and decide whether to mate with the male or to look elsewhere. The male generally doesn't help with reproduction beyond mating, that is, he doesn't help rear the chicks or build a nest and so the female likely gains only indirect benefits through choosing males with better genes for her offspring. Such can be the competition between bowerbirds that males sometimes sabo-tage each other's bowers, and even steal their decorations. The decorations males use often complement their own plumage colours, enhancing the range of coloration used in their combined displays. Richard Dawkins called structures found in nature such as bowers and their decorations 'extended phenotypes' because the animal's genes can also be reflected in the quality, skill, or some other attribute of the animal building the bower beyond its own physical appearance. In mate choice, extended pheno-types allow males to display aspects of their quality that cannot be demonstrated through the appearance of the male itself.

John Endler from Deakin University and Laura Kelley, now at Cambridge, have studied courtship behaviour in the great bowerbird (*Ptilonorhynchus nuchalis*) in Australia. Males build bowers consisting of a U-shaped structure made of twigs with high parallel walls (Figure 66).

FIG. 66. Top: A male great bowerbird (*Ptilonorhynchus nuchalis*) standing in front of his bower. Bottom: the grey and white decorations of the bowerbird, arranged according to a size gradient, with small objects closer to the bower structure.

Images Laura Kelley

The end of the bower is decorated with grey and white objects such as pebbles, bones, and shells, on which males display and sometimes place coloured objects. Females look through the long 'avenue' of the bower, 0.6–1 metre long, towards the other end where the ornaments are found. This is also where the male displays over his 'court', presenting his plumage and waving coloured objects in front of her. The high sides of the bower restrict the female's view so she effectively looks through a tunnel. Endler and colleagues initially showed that the male places the grey objects very specifically in a manner according to their size.[13] Specifically, he places smaller objects closer to the female's position, and larger objects further away beyond this. Remarkably, this creates a well-known visual illusion called 'forced perspective'. If you were to look along a tunnel with pebbles on the ground distributed randomly by size, then those pebbles further away would generally look smaller. This effect is useful in helping us to judge distance. In a forced perspective illusion, the gradient of increasing object size with distance cancels out this effect, creating the perception of a uniform arrangement of object sizes. The team conducted an experiment whereby they rearranged the objects on male bowers to destroy this illusion. The males responded by repositioning them, recreating the size gradient and thus suggesting that there is a benefit in mate choice to males in doing this. So what could that be? In essence, placing larger objects further away makes an area look smaller or shorter, the opposite of the *tromp l'oeil* effect of placing smaller objects further away from the viewer to give the sense of a scene extending into the distance, a trick often used in gardens. Making the display area look smaller may in turn make the male look bigger. The avenue itself serves to ensure that the female is forced to observe the male and the court only from an angle at which the illusion works.

Kelley and Endler next showed that males differ among one another in how effectively the forced perspective illusion is made (i.e. how well the size gradient is produced and maintained), suggesting that this could be a reliable feature for females to use when judging mate quality.[14] Consistent with this, the probability of mating success for males is higher when a female's view of the gradient is more effective.[15] That is, the stronger the illusion of even object sizes over distance, the higher the chance the female will mate with the male. However, things may not be quite so simple. Another recent study, by Natalie Doerr and John Endler, found that the ability of males to make illusions of different gradients was most influenced by the range of potential decorations males had available to use, rather than the specific male himself.[16] This suggests that the quality of the illusion a male can produce is affected by what objects are present in the environment as well as by his inherent sensory and cognitive skill at making the avenue. However, this does not preclude the possibility that better males are more capable of finding more useful ornaments, or that they hold territories with a better range of ornaments. So the quality of the illusion may still primarily relate to male quality, and indeed we know from Kelley and Endler's work that the illusion does seem to be important in guiding female choice and male success.

So what about the coloured objects males also present to females—what is their role in all of this? These are revealed and then hidden in turn by males from the end of the bower structure, and seem to grab and hold a female's attention and keep her at the bower for longer. Males often pick up and display coloured objects in turn, after which they drop or throw them across the avenue of the bower in front of the female. Such flashes of colour are enhanced by being presented against the grey background of objects on the ground, making the thrown ones appear more colourful. So the male bowerbirds essentially use two visual tricks that exploit the way vision works: to alter a female's perspective of size, and to enhance the vibrancy of the coloured objects males use in displays.

It is not clear how widely illusions are used by animals in nature, though various other birds, for example some South American mana-kins, also display against patches of ground that they clear of objects or manipulate in some other way. It is also likely that other animals have colour patches that make them appear larger than they really are. The perception of a male's coloration is also affected by other individuals around them. For instance, male guppy fish (like those often bought in aquarium shops), with their gaudy orange and blue body spots, will sometimes choose to surround themselves by less attractive individuals (with smaller colour patches) when they display to females.[17] This can create the impression that one male is better looking than he really is, making a mediocre male look attractive when surrounded by duds. It is also likely that animals create illusions of movement in nature. Indeed, we already encountered the idea of motion dazzle used by prey animals to mislead predators about their speed and direction. Recently, just like humans, fish have been shown to be fooled by the famous 'rotating snakes' illusion, in which a strong sense of movement is created in stationary spiral patterns of repeating patches of blue, yellow, and black colours.[18] This shows at least that other animals see illusions of movement where none in fact exist. So although we know very little about this subject yet, it seems almost certain that animals use illusions when displaying much more often than currently known.

As we know already, biases or preferences for certain stimuli, such as preferences of pollinators for certain flower colours, are widespread in animals, and such biases can be exploited during mate selection as well. Sometimes, preferences are learned, or at least reinforced with experience. However, they are also often innate and inherited, a natural part of the architecture of an animal's sensory and cognitive systems that govern how it behaves. For example, we know that of those primates equipped with good colour vision, many, including rhesus macaques and humans, have a preference for the colour red, and this can influence their mating decisions. When shown photos of males, female macaques prefer those surrounded by the colour red, as opposed to other colours.[19] Humans also show similar enhancement in mate attraction when the colour red is present, including on people's clothing. It is not clear why, but there is little doubt that the human and rhesus visual system is well tuned to detect red colours, probably because over evolution the ability to see red enabled our ancestors to pick out ripe red and yellow fruits against a backdrop of green leaves. This is something that most mammals cannot do effectively because these species lost the ability to discriminate between red and green colours early in their evolution, with selection favouring enhanced overall sensitivity to light and dark, an advantage in nocturnal conditions, over good colour vision. However, certain groups of mammals, including some primate groups, human ancestors among them, 're-evolved' effective colour vision. In primates, the capability to see red colours arose prior to the evolution of red skin and fur coloration, which happens to be widespread in species with good colour vision (Figure 67). Therefore, red mating signals apparently evolved to exploit

FIG. 67. A Japanese macaque (*Macaca fuscata*, left) and a bald uakari monkey (*Cacajao calvus*, right). Many primates have evolved red faces and other body parts, used in mate choice and dominance behaviour. This seems to have occurred after some groups of primates (including human ancestors) evolved a type of colour vision that enabled them to distinguish red, green, and yellow colours.

Left image Emma Manners/123RF; right image Jesse Kraft/123RF a pre-existing sensitivity or bias in primate vision that arose initially under selection for enhanced foraging success.[20]

Biases like those discussed above mean that females of numerous species possess 'latent' preferences just waiting to be exploited by males that subsequently evolve displays that target them. For example, Mike Ryan from the University of Texas, Austin (along with many students and colleagues), has spent several decades studying the mating behaviour of the túngara frog (*Engystomops pustulosus*) from Panama. Males call from aggregations in ponds, at which females approach and pick the individual with whom they want to mate. Ryan and others showed that the male call consists of two main components, a long 'whine' followed by one to seven short 'chuck' components (a 'whine-chuck' call).[21] Females prefer males that make both the whine and the chuck parts of the call, compared to those that only make the whine, and female hearing is well tuned to the frequencies of the chuck components. What's interesting is that in a related species of frog (*Physalaemus coloradorum*), males produce only a whine call, yet female hearing is also very sensitive to the túngara chuck components. In fact, females of this other species even prefer the calls of males of their own species when chuck components are added to their calls, compared to males without it. This tells us that the sensitivity of female hearing towards certain call components likely preceded that of the male chuck calls, which instead evolved under selection to exploit this sensory bias because it improved male mating success.

Similar research has found equivalent results in other animals. For instance, in swordtail fish, males of some species have an extension to part of their caudal (tail) fin, with a long sword-like expansion at the back. Females prefer males with the sword, especially longer ones, to those without, and the same is true for some closely related species in which males naturally lack swords, but adding one on artificially increases male mating success.[22] At this point it is important to highlight that when male signals exploit female sensory biases it does not necessarily mean that the females lose out or are in some way disadvantaged. In fact, we often don't know whether the heightened attraction of the females to the males' calls is in any way costly to them. Whether enhanced male calls actually trick females to mate with a male that they otherwise would not choose, like an inferior individual, is not clear. In fact, on the contrary, more elaborate displays often correlate with some favourable attribute such as size or strength, providing some benefits to the female. Strictly speaking then, this may not be truly deceptive in terms of being costly to females, although the signals do seem to have evolved to exaggerate female responses to them rather than simply conveying information about male quality.

Although exploitation of female preferences is likely to be widespread, it's not always clear why female biases exist and where they come from in the first place. However, one of the most common routes by which they are likely to arise is when male courtship displays resemble something females are highly responsive to but in another context, such as preferences for red during foraging in primates. Sometimes, male displays can even directly mimic something a female would encounter elsewhere. These so-called sensory traps are not well understood, but often involve the male's display or structure resembling something else and producing an out-of-context response in the female. The idea of a sensory trap was first suggested by John Christy, then at the Smithsonian Tropical Research Institute in Panama, back in the 1990s.[23] He argued that sensory traps do not just work by strongly stimulating an animal's sensory system (such as through loud sounds) but also include an aspect of mimicry, by which the signal resembles something else entirely and produces a strong response in the 'wrong' context. In the early 2000s, Christy and colleagues presented results from several studies on fiddler crabs (*Uca musica*) that supported this idea.[24] Males in many fiddler crab species build burrows and display to females during mating. In *U. musica*, males build hood-like structures over their burrows which females are attracted to. Crabs from other *Uca* species don't have these hoods, but Christy showed that the females are still attracted to them. The reason, it seems, is that they attract female attention and offer extra cover from predators by resembling other objects in the environment, bringing females to the males.

Christy's work was the first to describe properly the process of sensory traps, but it was not actually the first to show evidence for them, nor perhaps did it provide the best evidence for how sensory traps work. In the early 1990s Heather Proctor demonstrated that male water mites mimic the vibrations caused by their copepod prey during mating displays, and that these vibrations attract females.[25] Males walk or swim around until they encounter a female, at which point they vibrate their legs. This 'courtship trembling', as Proctor put it, induces females to respond to males initially just as they would when attacking prey. Not surprisingly, females are even more likely to mate when they have not eaten recently. As with the túngara frogs and swordtail fish, water mite hunting behaviour evolved before that of male courtship trembling, and so the male mating display exploits a pre-existing preference in females. Again, whether or not sensory traps carry a cost to females is not always clear. One might imagine that they often might though, especially if females are duped into repeatedly responding to courting males when all they really want to do is find food. Indeed, recent work on North American Goodeid fish, where males have yellow tail fin stripes resembling a damselfly larva, shows that the deception results in reduced foraging success in females because they are too often duped into nibbling at males instead.[26] Interestingly, females of some of these species appear to be evolving resistance to male displays, being better able to separate feeding and mating responses to the appropriate stimuli. This suggests that the cost is significant enough for female defences to evolve.

Sensory traps can sometimes take wonderfully unexpected pathways, not least in the courtship sounds made by some male moths. Ryo Nakano from the University of Tokyo and colleagues studied a species of moth called the common cutworm (*Spodoptera litura*).[27] In this species, males have special tymbal organs to produce ultrasonic sounds—the same type of organ that many moths use to deter bat predators (see Chapter 5). In moths like this and other species, female hearing is essentially already preadapted to detect male courtship songs because hearing organs in insects primarily evolved to detect incoming bats and their echolocation calls. Once ears arose, males could utilize this as a potential channel of communication during courtship, including in the cutworm moth. Nakano and colleagues showed that when males had their tymbal organs damaged, and were therefore unable to make sounds, their courtship success with females declined from around 95 per cent to 40 per cent compared to intact males. However, when the researchers accompanied the muted males with playbacks of mating calls from speakers, courtship success was restored. What's really interesting, however, is that this was also true when the team accompanied muted male moths with playback sounds of bat echolocation calls—success rate in courtship rose to between 91 and 100 per cent, depending on the type of bat call used. On the other hand, simply playing random noise from the speakers did not recover mating success. This seemingly rather odd finding makes sense when we consider how the male calls work and why. Female hearing is well tuned to the ultrasonic calls of bats, meaning that they have a sensory bias to detect these and other similar sounds that could occur in another context, including mating. Importantly, the male calls work by causing females to freeze. Essentially, they stop the females from running away, allowing the male to get close and mate. This freeze response likely also arose from a preadaptation to avoiding bats, because many insects freeze and drop to the ground when a bat is detected. So the male moth calls are both good at exploiting female hearing and related behavioural responses, both of which originally evolved to avoid bats.

Nakano and colleagues have also shown that males of another species of moth, the yellow peach moth (*Conogethes punctiferalis*), use calls that resemble bats to deter rivals as well.[28] They showed that males produce two types of call, one comprising a long pulse and the other several short pulses. The long call induces females, who emit pheromones to draw in males, to take up a mating posture. However, to mate successfully males must also beat off competition from rivals. The short pulses seem to resemble the echolocation calls of attacking bats, manipulating rival males to stop approaching and remain still. Indeed, both moth pulse calls and bat echolocation calls have this effect. In this instance, the deception clearly carries a cost to other males through lost mating opportunities.

Exploiting the sensory and cognitive systems of potential partners is one way to increase mating success, but in nature even more outlandish methods exist. Because males often have to compete with one another for access to females, competition is frequently intense. The weapons and strength that males of many species possess, including horns and antlers on many deer, large canine teeth in certain primates, or the sheer size and weight of elephant seals, are testament to this intense selection pressure. Competition can be especially fierce when a select few males are able to aggressively defend multiple females from rivals, enabling them to monopolize mating opportunities. For individuals that are weaker, smaller, or have inferior weapons this is bad news because their reproductive potential can be substantially diminished. So what other options do they have? Well, weaker males could become sneaky, and rather than fighting with the competition try to evade it, darting in to quickly mate with a female before rushing off again. For example, in elephant seals the huge alpha male defends a harem of females on the beach, to which he tries to monopolize access. In theory, smaller males could do better by lurking round the edges of the harem and then dashing in while the male is not looking.

Being a 'sneaker' male may be good enough, but such males still run the risk of aggression and attack from the bigger individuals, and their chances of success are still limited by opportunities when the dominant males' backs are turned. Better still would be to find a way of being sneaky in plain sight, without other males even knowing. This is precisely what happens in quite a few animals, in which some males of the species mimic the appearances of females. Bluegill sunfish (*Lepomis macrochirus*) are a species relatively well studied for female mimicry since the late 1970s. They are found in the lakes of North America. Dominant males aggressively defend nesting territories, sometimes within spectacular breeding colonies of up to 150 individuals. Females are attracted to these territories, where they spawn.

Wallace Dominey from Cornell University showed in 1980 that territorial males from a lake in New York tend to be more successful when they are larger.[29] However, also existing within the population are female mimics, which are smaller than the territorial males and rarely engage in aggressive interactions. They resemble females in their dark colour patterns and approach males in their territories. If things work out they add themselves on to a spawning pair of a territorial male and real female, and while the male thinks he's mating with two females, the mimetic male is in fact also releasing sperm and trying to fertilize the female's eggs, which the territorial male then cares for. The female-mimicking males may be smaller than the territorial males overall, but importantly one part of their body is relatively bigger: the testes. This enables them to produce lots of sperm to enhance the success of these tactics. Dominey also showed that there was no substantial difference in the age distributions of mimetic males and territorial individuals, suggesting that this strategy is a fixed way of life for those individuals, rather than a stage of development that all males must pass through.

Dominey's work suggested that there were two types of fixed male sunfish: a female mimic and a nesting male. However, later that same year Mart Gross and Eric Charnov from the University of Utah showed that the situation was more complex, at least at their study site in Canada.[30] They showed that there were, in fact, three types of male in a population: the normal nesting males, in addition to both sneaker and 'satellite' males. The satellite males were most similar to those Dominey had identified, looking like females and hanging around the sites to engage with a mating pair. Sneaker males, however, looked more like normal males but rapidly darted into and out of male territories, using speed instead of deception to release sperm. Gross and Charnov showed that being a dominant male was, as Dominey had thought, a strategy only employed by some individuals, and arose when males delayed maturation until they were about seven years old. In contrast, sneaker and satellite males were the same individuals separated in time; they start off at around age two as a sneaker, and then develop at around four to five years into a female-mimicking satellite male. Sneaker males gain success through rapid movement and stealth, whereas female-mimicking satellite males achieve success through directly tricking the nesting male into thinking it's a female.

The 'normal' males and female mimics or sneaks represent alternative reproductive strategies. Such alternative strategies have long interested evolutionary biologists because we expect interesting issues to arise with regard to the frequency of these strategies over time. Female mimics, for example, shouldn't exist alone because there would be little point in behaving in that way if every other male was a mimic too; a male would do much better to defend access to females instead and drive away other individuals. On the other hand, when female mimics are rare, then 'normal' males can let their guard down because the chances of being duped are low, such that there should be a high pay-off to being a mimic. So we may expect alternative mating strategies to either reach some sort of stable equilibrium, in which the pay-offs to both strategies are about the same (i.e. changing strategy from being a mimic to a normal male would not help and the male would likely do worse, and vice versa). In evolutionary biology, this is often called an *evolutionarily stable strategy*, because nobody can do better by switching roles to something else. This does not mean that we should end up with equal proportions of each strategy in the population, because one strategy may be more successful in itself, all things being equal. At the moment it isn't clear whether this does explain most alternative mating strategies, but it is likely to be the case in at least some animals. In contrast, in many cases different strategies arise depending on both the age or condition of an individual and the make-up of a population as they develop. For example, a relatively small male in a population of dominant individuals should develop into a sneaker male as it matures. For example, in horned beetles (*Onthophagus acuminatus*) the strategy a male employs depends on nutrients acquired during development. Larger, more nourished males develop full horns that they use to defend the entrance of a tunnel containing a female. However, smaller males don't develop proper horns and instead become sneaks. They dig a new burrow that bypasses the entrance guarded by the male to intercept the female elsewhere.

In general, female mimicry and sneaker strategies appear to be quite common in fish, perhaps partly because many species spawn their eggs and sperm directly into the environment or on to the substrate, making it easy for another male to dart in and release his sperm at the same time. In fact, sneaking has been reported in at least 140 fish species that have external fertilization.[31] How successful sneaker and female-mimicking strategies are compared to 'normal' males is unclear, but the strategy has seemingly evolved multiple times in fish. It is difficult to assess the benefit of these strategies because it does not just depend on the relative number of fertilizations a male gets in a breeding season, but rather his lifetime success. Furthermore, female mimics could benefit if they do not have to invest energy and nutrients developing elaborate and costly ornaments or displays to attract females, or simply the time and energy trying to monopolize access to females. In effect, sneakers and mimics can save energy and investment for other tasks. And the duped males in the population could fight back through greater aggression towards sneakers. Determining the relative success of each strategy presents a challenge for researchers.

Female mimicry is by no means restricted to fish, but also occurs in other animal groups. Very much like the bluegill sunfish, a species of bird called the ruff (*Philomachus pugnax*), a type of sandpiper, has three distinct male strategies.[32] The males are rather odd-looking birds, with fluffy collar-like ornament of feathers just like the Elizabethan ruffs that inspired the species name. Ruffs comprise males in breeding plumage that defend small territories, and satellite males that don't mimic females but rather steal copulations with females when the resident males are distracted. The third morph is the only clear example of a male bird that mimics females for mating success. They behave as sneaker males, tricking the resident males into failing to realize they are a threat. Like the equivalent bluegill strategy, sneakers also have enlarged testes to produce lots of sperm. Unlike the horned beetles though, a father's strategy is passed on to male young, meaning that mating strategies are not simply due to the condition of the males but rather are genetically fixed.

Female mimicry occurs in invertebrates too. For example, in the Australian giant cuttlefish (aptly named as individuals can reach six kilograms in weight) thousands of individuals aggregate in areas off the south coast of Australia in the winter to spawn. In these aggregations, males often outnumber females by several individuals to one, resulting in intense male–male competition. Unsurprisingly, certain individuals resort to sneaker strategies and to female mimicry.[33] Some males lurk in the open water in full view waiting for the male of a pair to become distracted before making a move and mating with the female, whereas others do much the same but from the cover of a hiding place. Cuttlefish are also famed for their colour-changing skills (as we know from their camouflage, see Chapter 4), and some small males change coloration to mimic females. In fact, this sometimes results in males trying to mate with them too.

Finally, sexual mimicry is not always restricted to males. In several damselfly species, females exist in morphs of different appearance, one of which looks like a male. There are various possibilities for why this evolved, including that males are choosy in mate selection. However, in many species, such as *Ceriagrion tenellum*, a relatively small red-coloured damselfly found in Spain, this does not appear to be correct because female mating success does not seem to be influenced by morph type.[34] However, males are often confused by the female morph and mistake it for a male, and are more attracted to female-like morphs. The answer, it seems, relates to preventing male harassment. Females that mimic males can avoid some of the unwanted attention of males trying to mate with them, and consistent with this is work showing that male-like morphs are more common in *C. tenellum* populations with greater densities of males. Reduced harassment is likely to be beneficial as, unlike males, many females only want to mate a limited number of times and harassment can disrupt other activities (like foraging) or even result in physical damage. The cost to females, however, could arise in reduced opportunities to mate when the female actually wants to, in particular when the population density is low. So, overall, the stable outcome is that female success is often about the same for both the 'normal' females and male mimics, but the proportion of these depends on the density of males in the population. Intriguingly, females of some damselfly species may have come up with a great solution to balance mating opportunities and avoid male harassment. The common bluetail damselfly (*Ischnura heterosticta*) is an Australian species usually found in vegetation by still-water habitats such as lakes and ponds. Males are a stunning bright blue but females come in a variety of forms. Scientists have found that sexually immature females mimic the blue coloration of males, and that this reduces unwanted harassment and allows them to get on with other things, such as foraging for food. However, when females mature and are ready to mate they irreversibly change colour over the course of twenty-four hours, from a bright blue to a greenish-grey, after which they look unlike the males (Figure 68).[35] While male mimicry is most widely studied in damselflies, some female squid also adopt colour patterns that perhaps mimic the attributes of males, and again this might enable them to reduce harassment by unwanted male suitors.[36] Beyond these examples, little is known regarding how common male mimicry is by females in nature, or whether it always functions to prevent harassment.

In evolutionary terms, the success of an individual relates substantially to how effectively they pass on their genes over their lifetime and

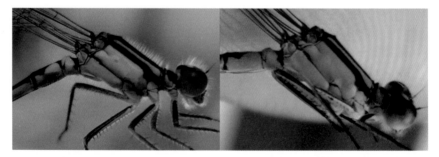

FIG. 68. A female damselfly (*Ischnura heterosticta*) that changes colour from a male mimic into the reproductively active female form in approximately twenty-four hours.

Images Shao-Chang Huang

how they may ensure the subsequent success of their offspring. But the interests of different parties in this process frequently conflict, whether it is plants minimizing the reward to pollinators, or males trying to coerce reluctant females. This has resulted in a plethora of examples of deception and manipulation, ranging from exploiting the way that animal sensory systems work, through to elaborate mimicry of other objects. Once again, these examples reveal much about how communication signals (honest and deceptive) are tuned to the way that animals perceive the world, and also the various processes and mechanisms that underlie evolution and adaptation. This includes aspects of frequency-dependent selection, and also of how the targeted parties can again fight back against cheats, either over evolution or during their own lives through learning. In years to come, there is little doubt that our understanding of sexual deception will increase substantially. We have barely scratched the surface on many topics, from sensory traps to the use of illusions.

9

THE FUTURE OF DECEPTION

· · · · · · · · · · · · ·

The idea that animals trick and cheat one another is far from new. As I have tried to highlight throughout this book, many of the initial ideas surrounding deception were conceived by early evolutionists and pioneering Victorian naturalist-explorers, including the likes of Wallace, Bates, Poulton, and Darwin himself. These were individuals with an abundance of insight and knowledge about natural history, so it is perhaps unsurprising that they paved the way on this subject, just as they did with many others. What might be less expected is the number of theories and examples that have been tested effectively in the last decade or so. Despite much early work, stemming back to Wallace, Bates, and others, it is only recently that studies of many areas of deception have moved from largely descriptive and anecdotal accounts to rigorous scientific experiments, of which I've tried to give a flavour throughout. A striking number of advances have been made recently. It is evident that deception in nature remains a vibrant area of research today, in part because it's a fascinating subject in its own right, but also because it is widely used as a system to understand fundamental areas of biology, ranging from how organisms communicate to how evolution works. This is true both in terms of understanding the mechanisms governing these areas (for instance the molecular and

genetic basis of traits), through to the large-scale processes that underlie evolutionary change at the level of entire populations. The study of
deception is also a good way to understand how sensory systems work,
not least because many forms of deception have evolved under selection to exploit the sensory (and cognitive) processing of those animals
they exploit. No doubt increasing understanding of sensory systems
has also fed back into the relative surge in research on deception. In
addition, research on deception has been aided by significant advances
in technology, including devices to measure the colours, smells, or
sounds of animals and to understand and model how animals perceive
the world, through to clever methods to create carefully designed
stimuli in behavioural experiments. In many cases, the research
I have described is quite simple in principle, but the methods used to
make the stimuli and to analyse the animal responses are sophisticated.

While many of the original ideas regarding deception are old, the
strategy itself is ancient. For example, although rare, fossils have been
found from as far back as the early Cretaceous (around 126 million
years ago) showing insects mimicking leaves to provide camouflage.[1]
We might speculate that masquerade began to evolve with the advent
of early insectivorous birds and mammals and quickly became specialist in function and appearance. Much more ancient, it has been suggested that some Cambrian (from around 500 million years ago)
marine brachiopods may have evolved Batesian mimicry to resemble
unpalatable sponges that seem to have existed at about the same time.[2]
Ant mimicry by spiders has also been reported as occurring at least fifty
million years ago from specimens preserved in amber.[3] Beyond
defence, the luring of prey by predators is also ancient: anglerfish, for
example, are thought to have arisen around 145–170 million years ago
and diversified from there on.[4] As more fossils are discovered, hopefully other types of deception will be unearthed too, shedding light on
their evolution through time.

Many deceptive strategies are remarkably sophisticated, including forms of mimicry and camouflage. So how did they evolve over time? In the case of processes like sensory exploitation, it is not difficult to imagine that loud or bright signals, which are very effective at stimulating other animals' sensory systems, would be more successful and continue to be enhanced over evolution. Likewise, we could also imagine how something like an ancestral hoverfly might through chance mutation gain a slightly yellower coloration that is enough to put a hint of doubt in some predators' minds that the hoverfly might actually be a wasp or bee. But this is largely just storytelling, and how can we explain sophisticated examples like species that intricately resemble the form, colour, and shape of dead leaves and twigs? How could that type of deception first begin to evolve given that a close match to a real leaf underlies how it works? This question takes us back to a debate that involved the very first evolutionists, whereby some (like Darwin) argued that evolution primarily works on small incremental changes, whereas others (like the US geneticist Richard Goldschmidt in the 1940s) favoured the idea that large-scale changes happened suddenly, and with few intermediates (sometimes called 'hopeful monsters' waiting to arise).[5] Wallace also believed that leaf-mimicking butterflies and stick insects evolved their deception over an 'immense series of generations'.[6] In truth, in many species we don't know for sure, but in 2014 research was published by Takao Suzuki and colleagues from the National Institute of Agrobiological Sciences in Japan showing that elaborate deception can, and at least sometimes does, evolve via small incremental changes in form.[7]

The work by Suzuki and others was, rather nicely, undertaken on the leaf-mimicking butterflies *Kallima* from South East Asia, the species that so impressed Wallace and others and that helped to provide early examples and evidence of evolution (see Chapter 4). The study analysed the shape and morphology of existing *Kallima* species, and of various related butterflies, coupled with a family tree derived from molecular genetics. With this they could deduce the patterns of relatedness among species. In the first instance, they determined that all the butterflies from the group shared the same basic ground plan (a bit like a fundamental blueprint across all species), and that components of this ground plan had been modified over evolutionary time from their common ancestors into a number of different states encoding appearance (such as wing veins, spots, and other markings) (Figure 69). The team also worked out which features were likely to have been possessed by the ancestral butterflies, and then looked at how *Kallima* has changed since, including traits that seem to mimic leaf veins and small spots of mould. They found that leaf mimicry has evolved through a series of intermediate forms, each one a better match to a real leaf than the previous. This finding shows that deception in *Kallima* did likely arise through comparatively small incremental changes, just as Darwin and Wallace would have thought. While a few scientists may disagree, I personally find this suggested pattern of evolution for even complex forms easy to imagine. It's not hard to see how an ancestral butterfly that, say, evolved brown coloration would blend in with dead leaves better than individuals that were less brown, and subsequently how another ancestor developed a crude leaf shape, further gaining a slight but important survival advantage. Given sufficient time (and remember, we know that masquerade is at least 126 million years old), the disguise could become extremely sophisticated. Of course, this is just one example. Whether other elaborate forms of deception have arisen by small gradual steps or comparative leaps of evolution remains to be seen. My feeling, given the widespread presence of imperfect mimics and what we know about how mimicry does not always have to be perfect to work, is that gradual change will be very common. Indeed, Suzuki has also recently shown that a moth called *Oraesia excavata*, from the huge family Noctuidae, has also evolved its

FIG. 69. The leaf-mimicking butterfly *Kallima inachus* and how changes to the butterfly 'ground plan' have underpinned its evolution. The top image shows *K. inachus* with the key elements of its ground plan highlighted. The bottom image shows the analogous components of the ground plan of a wide range of butterfly species (here principally Junoniini and Kalimini), which have been modified over evolutionary time from their common ancestor. In the case of *Kallima* this has been modified to produce increasingly effective leaf mimicry through a series of gradual changes.

Images Takao K. Suzuki and modified from Suzuki et al. (2014)
BMC Evolutionary Biology, 14:229

mimicry of leaves through changes from its ancestors, again involving a modification of a general ground plan found in butterflies and moths.

So what do all the various examples of deception we have covered tell us about how animals and plants fool one another? First, we shouldn't think of animals that are tricked as passive observers. They often fight back, with individuals either learning to be cautious of cheats, or populations evolving inherited ('innate') defences over time. This means that many deceptive systems are really quite dynamic, either in real time or over evolution. For example, in drongos, the birds that make false predator alarm calls to steal food from group-living birds and mammals (see Chapter 2), individual drongos do not just make false alarm calls all the time. Instead, they must also make some genuinely honest alarm calls when predators are nearby. Otherwise the animals they are tricking would learn to ignore them. Relationships between deceptive individuals and those they cheat should also vary at the population level. It is widely assumed in Batesian mimicry that the mimics, such as a hoverfly, impose costs on the model, such as a wasp. This is because the animal the hoverflies are tricking, a predator, may after attacking a hoverfly learn incorrectly that insects with black and yellow stripes are actually harmless. Subsequently, they can end up attacking and killing wasps too. As a result, we expect the frequency of the hoverflies to be relatively low and linked to the frequency of the wasps so that the trickery and defence still work. Beyond this, we also predict that individuals of the species being exploited might come under selection to change their appearance (over generations) in order to try and 'escape' mimicry. In doing so, the effectiveness of the mimicry is lessened and the cost to the model is reduced. This can lead to an evolutionary chase, whereby the mimic also has to keep changing in order to retain its function. Indeed, there is evidence that this does occur in different systems, including recent evidence of changes in the egg colour of the parasitic cuckoo finch and its host we discussed in Chapter 7,[8] and in Batesian mimicry by salamanders.[9]

Second, deception can teach us how our perception of the natural world can frequently be inadequate, and at times simply wrong. The sensory systems of any animal do not encode all the potential information in the world around them—there's just too much of it. Instead, they evolve senses that encode the most relevant stimuli used to carry out their daily lives. Cats have a highly refined sense of smell, far superior to ours, enabling them to detect and identify individuals that were present in a given location hours ago, an ability valuable for things such as marking and identifying territories and finding prey. Their night vision is also excellent, with high sensitivity enabling them to remain active and hunt during darkness. Yet their colour vision is poor compared to ours, restricted to a limited range of colours that we would see as blues and yellows. As with many other mammals, cats cannot, for instance, distinguish between colours that we see as red, green, and yellow, perceiving them all as the same colour type. Cats are not thought to have a particularly impressive sense of taste either. In short, their perceived world is a product of the sensory systems they have, which have evolved under different selection pressures, and it is different from ours. In addition, constraints on the way that sensory systems are made can also affect perception. For example, the compound eyes of many invertebrates, with their thousands of tiny lenses, can create an image of the world that is somewhat coarse or pixelated, and lacking in fine detail. This limits many species' ability to see small patterns and intricate shapes. On the other hand, male moths are often capable of detecting minuscule concentrations of female pheromones in the air, using that information to orientate themselves towards potential mates.

There are countless other examples I could give, but the point is that the senses of any animal are tuned by evolution to pick up information from the world around them that has been most relevant to their ancestors. They may be detecting only a snapshot of what other species might perceive. Humans are no exception. This means that deception takes place in a way that is most salient to the animals being deceived, with regard to their sensory apparatus, and using our own perceptions to judge this can be misleading, either missing the sophistication of deception because we don't perceive it properly, or even perhaps thinking the deception is not very good because it arises in areas in which our senses are superior to those of the animals being tricked. At times, we might even completely misunderstand what's going on. If we return to the Australian crab spiders that lure insect prey to the flowers they sit on (see Chapter 1), to us the spiders often look wonderfully well camouflaged. In fact they glow like beacons in UV light, which their prey can see. Only with specialist equipment can we reveal that the deception works not by the spider lying hidden and waiting for its victims, but instead by luring them towards an attractive colour. Some of the orchids we encountered in Chapter 8, which deceive male wasps through sexual mimicry, illustrate the opposite point. Here, the orchids to our eyes have only a passing resemblance to a female wasp, yet to the compound eyes of the male wasp, which also lacks an effective ability to see red colours, both the orchid and real wasps are almost indistinguishable. As a result, we have probably only begun to scratch the surface regarding how widespread deception is in nature and how it works. Much of what we know about deception comes from examples in the visual sense, and to a lesser extent smell and hearing. How widespread it is in other senses remains to be seen. Indeed, some animals have entire sensory modalities that we lack. For example, groups of electric fish from Africa and South America produce and detect weak electric signals from specially modified organs, and use this ability for navigation, hunting, rivalry, and mate choice. I would not be surprised to find electric deception existing among them.

We can also apply these considerations to conundrums such as the numerous cases of imperfect mimicry in nature, in which a supposed mimic does not resemble its model especially closely. As we noted, there are many potential explanations for this, including that evolution may simply not have had the time as yet to perfect the level of mimicry—it is a work in progress. But sometimes it might simply be the case that the mimicry is good enough to work well. To the sensory system of the animal being deceived, it might actually be quite effective. Furthermore, some cases of imperfect mimicry are likely to exist because the animal being tricked only pays attention to certain features and not others. For example, hosts of the African cuckoo finch primarily reject foreign eggs based on which features of egg pattern and colour are most likely to give the game away in terms of whose eggs belong to whom. Likewise, laboratory experiments with birds such as pigeons show that they focus their efforts on learning the most salient information to define different types of object (e.g. wasps or flies), and use this to categorize new objects (e.g. hoverflies) later on. These cases suggest that effective mimicry can work even when only a limited set of traits are mimicked (e.g. colour but not pattern) if the animal being tricked does not pay attention to other traits.

Undoubtedly, one of the major challenges for future work on deception is to better understand how exactly it works. Specifically, we need to determine when processes such as sensory exploitation are involved, that is, when an organism makes use of pre-existing sensory or even cognitive preferences of an animal, as opposed to when genuine cases of mimicry occur, when an animal mistakenly categorizes something as the wrong type of object. There are unambiguous examples of both these processes operating in deception. For example, the Australian crab spiders clearly lure pollinator prey that have a general preference for UV colours. In contrast, it's hard to argue against the jumping spiders that resemble ants being a genuine case of mimicry. The difficulty comes, however, with all those instances where we can be less sure about what the mechanisms involved are. For example, does the lure on an anglerfish really mimic a prey item to its victims, or does it just exploit a general preference for small moving objects? To make matters more complex, sensory exploitation and mimicry are not mutually exclusive. Indeed, the yellow wing patches that Japanese hawk-cuckoo chicks flash at their foster parents illustrate this. The parents must, at least sometimes, consider the yellow markings as a real mouth because they sometimes try to feed the wings, yet the appearance of each wing marking is brighter and richer in UV light then any real gaping chick. It is what scientists often call a supernormal stimulus, eliciting a heightened response in the target animal through its extreme properties.

You would be forgiven for wondering whether all this is just semantics—scientists simply arguing over what to call something. However, it does matter because if we are to understand how deception arises in different species, by what pathways it evolves, and where it might ultimately end up and the form it might eventually take, we need to know how it works. To do so means designing studies that test how other animals categorize objects. The work we looked at on twig mimicry (masquerade) by caterpillars against naive chicks (Chapter 4), or of using pigeons to categorize different hoverflies as either wasps or flies (Chapter 5), shows how this can be done, but these experiments are challenging. Sadly, we can't simply ask other animals to explain why they responded to an object in a particular way.

A final question we might ask is whether some animal and plant groups are more inclined to deceive than others, and if so why. On the one hand, deception does seem prevalent in certain groups, especially spiders and other predators that sit and wait for prey (or actively lure victims). Beyond this, deception seems to arise simply when there is a communication system of other species that is open to exploitation, and when the selection pressure is sufficiently strong for deception to be worth the risk of being caught out (e.g. when avoiding predation or securing a mate). Perhaps a more important question is to ask why certain animals evolve one type of deception (e.g. Batesian mimicry) and others a different type (e.g. camouflage). What drives evolution down these different pathways? Certainly the life history of a species—where it lives, how active it is, what it feeds on, and so on—matter greatly, but at the moment we don't know a great deal about when and why one form of trickery is favoured over another. Some researchers have suggested that particular regions of the planet, specifically Australia, might have more deceptive species than other locations.[10] This suggestion is based on the fact that if you count the number of known species that use deception belonging to three groups (spiders, avian cuckoos, and orchids), Australia seems to have a disproportionate number of them (compared to overall species numbers worldwide). For example, crab spiders in Europe seem to rely on camouflage and avoiding detection to hide from their prey, whereas Australian species often use UV signals to deceptively lure their victims. Australia also seems to have higher proportions of cuckoos and orchids than elsewhere. This is certainly an intriguing idea, and not implausible given the unique nature of other Australian wildlife. However, it might just be that Australia, Europe, and North America have been more intensively studied than elsewhere, and that deception could be especially rife in the complex yet vastly understudied tropical biota of South America, Africa, and South East Asia, where specialist and extreme modes of life are common. Interestingly, Wallace also argued that the comparatively stable environment and climate of the tropics, as opposed to temperate regions of the planet, might have allowed some of the most remarkable instances of deception to arise, in part owing to the large amount of time during which they have been able to evolve.[6]

Our understanding of deception has come a very long way, perhaps more than ever in the last decade. Deception has provided a suite of examples supporting and greatly enhancing our understanding of evolution and adaptation, and the ideas of early scientists and naturalists.

Much of this is thanks to the ingenuity of the wide range of scientific approaches used, from 'traditional' fieldwork and experiments, to sophisticated molecular genetics and modelling approaches to understand animal senses. In recent times, fundamental natural history and exploration of the diversity of nature has often been put aside to make way for new subjects and technological advances. Many biologists spend much of their working lives in the lab, rarely studying species in the wild. Old-fashioned natural history has fallen out of favour, and is even looked down upon by some. It's a far cry from the approach of Wallace, Bates, and their compatriots. Yet, as I have tried to highlight, natural history is where many of our great ideas about the workings of nature and evolution come from. Unless we understand the ecology and environments in which animals live, and the selection pressures they face, we risk losing touch with the species we seek to understand, especially at a time of dramatic loss of natural ecosystems. Thankfully, there is now a steadily growing return to incorporating natural history back into biological research, coupled with rigorous and pioneering experimental approaches. The study of deception is one such field. Perhaps what is most exciting is that there are many issues and conundrums still to resolve about how deception works and how it evolves. Wallace argued that animal coloration, including things such as camouflage and mimicry, offered 'an almost unworked and inexhaustible field of discovery for the zoologist'.[6] He wasn't wrong. Comprehending these and other aspects of deception will greatly enhance our knowledge of the natural world and the mechanisms that make it work, and continue to heighten our wonder at the diversity of life, and the remarkable power of evolution.

NOTES AND REFERENCES

1. Much work on deception by *Maculinea* butterflies is reported as having been undertaken on the mountain alcon blue (*M. rebeli*). However, David Nash from the University of Copenhagen, an expert on *Maculinea*, tells me that most work has probably been undertaken on the alcon blue *M. alcon*. True mountain alcon blue butterflies seem confined to a few areas 1,000 metres or more above sea level. See: Tartally, A., Koschuh, A., and Varga. Z, (2014) 'The re-discovered *Maculinea rebeli* (Hirschke, 1904): host ant usage, parasitoid and initial food plant around the type locality with taxonomical aspects (Lepidoptera, Lycaenidae)'. *ZooKeys* 406: 25–40.

   To add further uncertainty, some scientists have proposed that species in the *Maculinea* genus should be renamed to the sister genus *Phengaris* (e.g. *Phengaris alcon*). Not everyone agrees, and at the time of writing the case before the International Commission for Zoological Nomenclature is unresolved.
2. Around 75 per cent of all Lycaenid species seem to associate with ants. Most relationships are mutualistic, but around 200 species of these butterflies (4 per cent) are thought to be parasitic. Als, T.D., Vila, R., Kandul, N.P., et al. (2004) 'The evolution of alternative parasitic life histories in large blue butterflies'. *Nature* 432: 386–90.
3. Akino, T., Knapp, J.J., Thomas, J.A., and Elmes, G.W. (1999) 'Chemical mimicry and host specificity in the butterfly *Maculinea rebeli*, a social parasite of Myrmica ant colonies'. *Proceedings of the Royal Society B: Biological Sciences* 266: 1419–26.
4. Nash, D.R., Als, T.D., Maile, R., Jones, G.R., and Boomsma, J.J. (2008) 'A mosaic of chemical coevolution in a large blue butterfly'. *Science* 319: 88–90.
5. Barbero, F., Bonelli, S., Thomas, J.A., Balletto, E., and Schonrogge, K. (2009) 'Acoustical mimicry in a predatory social parasite of ants'. *Journal of Experimental Biology* 212: 4084–90; and Barbero, F., Thomas, J.A., Bonelli, S., Balletto, E., and Schonrogge, K. (2009) 'Queen ants make distinctive sounds that are mimicked by a butterfly social parasite'. *Science* 323: 782–5.
6. Schaefer, H.M. and Ruxton, G.D. (2009) 'Deception in plants: mimicry or perceptual exploitation?' *Trends in Ecology & Evolution* 24: 676–85.
7. Thomas, J.A. and Settele, J. (2004) 'Butterfly mimics of ants'. *Nature* 432: 283–4.
8. Heiling, A.M., Cheng, K., Chittka, L., Goeth, A. and Herberstein, M.E. (2005) 'The role of UV in crab spider signals: effects on perception by prey and predators'. *Journal of*

Experimental Biology 208: 3925–31;and Heiling, A.M., Herberstein, M.E., and Chittka, L. (2003) 'Crab-spiders manipulate flower signals'. *Nature* 421: 334.

9. Heiling, A.M. and Herberstein, M.E. (2004) 'Predator–prey coevolution: Australian native bees avoid their spider predators'. *Biology Letters* 271: S196–S198.

10. See various chapters in: Smith, C.H. and Beccaloni, G., editors, (2008) *Natural Selection and Beyond: The Intellectual Legacy of Alfred Russel Wallace*. Oxford: Oxford University Press.

11. Wallace, A.R. (1867) 'Mimicry and other protective resemblances among animals'. *Westminster Review* (London ed.) 1 July: 1–43.

CHAPTER 2

1. Munn, C.A. (1986) 'Birds that "cry wolf"'. *Nature* 319: 143–4.

2. Flower, T. (2011) 'Fork-tailed drongos use deceptive mimicked alarm calls to steal food'. *Proceedings of the Royal Society of London B: Biological Sciences* 278: 1548–55.

3. Flower, T., Child, M.F., and Ridley, A.R. (2013) 'The ecological economics of klepto-parasitism: pay-offs from self-foraging versus kleptoparasitism'. *Journal of Animal Ecology* 82: 245–55.

4. Cheney, K.L., Grutter, A.S., Blomberg, S.P., and Marshall, N.J. (2009) 'Blue and yellow signal cleaning behaviour in coral reef fishes'. *Current Biology* 19: 1283–7.

5. Côté, I.M. and Cheney, K.L. (2005) 'Choosing when to be a cleaner fish mimic'. *Nature* 433: 211–12; and Cheney, K.L., Grutter, A.S., and Marshall, N.J. (2008) 'Facultative mimicry: cues for colour change and colour accuracy in a coral reef fish'. *Proceedings of the Royal Society B: Biological Sciences* 275: 117–22.

6. Cheney, K.L. and Côté, I.M. (2005) 'Frequency-dependent success of aggressive mimics in a cleaning symbiosis'. *Proceedings of the Royal Society B: Biological Sciences* 272: 2635–6.

7. Cheney, K.L. (2012) 'Cleaner wrasse mimics inflict higher costs on their models when they are more aggressive towards signal receivers'. *Biology Letters* 8: 10–12.

8. Cheney, K.L. and Côté, I.M. (2007) 'Aggressive mimics profit from a model-signal receiver mutualism'. *Proceedings of the Royal Society B: Biological Sciences* 274: 2087–91.

9. Widder, E.A. (1998) 'A predatory use of counterillumination by the squaloid shark, *Isistius brasiliensis*'. *Environmental Biology of Fishes* 53: 267–73.

10. Hafernik, J. and Saul-Gershenz, L.S. (2000) 'Beetle larvae cooperate to mimic bees'. *Nature* 405: 35; and Saul-Gershenz, L.S. and Millar, J.G. (2006) 'Phoretic nest parasites use sexual deception to obtain transport to their host's nest'. *Proceedings of the National Academy of Sciences of the USA* 103: 14039–44.

11. Vereecken, N.J. and Mahe, G. (2007) 'Larval aggregations of the blister beetle *Stenoria analis* (Schaum) (Coleoptera: Meloidae) sexually deceive patrolling males of their host, the solitary bee *Colletes hederae* Schmidt & Westrich (Hymenoptera: Colletidae)'. *International Journal of Entomology* 43: 493–6.

12. Cortesi, F., Feeney, W.E., Ferrari, M.C.O., et al. (2015) 'Phenotypic plasticity confers multiple fitness benefits to a mimic'. *Current Biology* 25: 1–6.

13. Wallace, A.R. (1877) 'The colours of animals and plants'. *Macmillan's Magazine* September and October: 384–471.

14. O'Hanlon, J.C. (2014) 'The roles of colour and shape in pollinator deception in the orchid mantis *Hymenopus coronatus*'. *Ethology* 120: 1–10; O'Hanlon, J.C. (2014) 'Predatory pollinator deception: does the orchid mantis resemble a model species?' *Current Zoology* 60: 90–103; and O'Hanlon, J.C., Holwell, G.I., and Herberstein, M.E. (2014) 'Pollinator deception in the orchid mantis'. *American Naturalist* 183: 126–32.

CHAPTER 3

1. Herberstein, M.E., Craig, C.L., Coddington, J.A., and Elgar, M.A. (2000) 'The functional significance of silk decorations of orb-web spiders: a critical review of the empirical evidence'. *Biological Reviews* 75: 649–69.
2. Cheng, R.-C. and Tso, I.-M. (2007) 'Signaling by decorating webs: luring prey or deterring predators?' *Behavioral Ecology* 18: 1085–91.
3. Blamires, S.J., Hochuli, D.F., and Thompson, M.B. (2008) 'Why cross the web: decoration spectral properties and prey capture in an orb spider (*Argiope keyserlingi*) web'. *Biological Journal of the Linnean Society* 94: 221–9.
4. Cheng, R.-C., Yang, E.-C., Lin, C.-P., Herberstein, M.E., and Tso, I.-M. (2010) 'Insect form vision as one potential shaping force of spider web decoration design'. *Journal of Experimental Biology* 213: 759–68.
5. Tan, E.J. and Li, D. (2009) 'Detritus decorations of an orb-weaving spider, *Cyclosa mulmeinensis* (Thorell): for food or camouflage?' *Journal of Experimental Biology* 212: 1832–9.
6. Tan, E.J., Seah, S.W.H., Yap, L.-M.Y.L., Goh, P.M., Gan, W., et al. (2010) 'Why do orb-weaving spiders (*Cyclosa ginnaga*) decorate their webs with silk spirals and plant detritus?' *Animal Behaviour* 79: 179–86.
7. Bruce, M.J., Heiling, A.M., and Herberstein, M.E. (2005) 'Spider signals: are web decorations visible to birds and bees?' *Biology Letters* 1: 299–302.
8. Craig, C.L. and Ebert, K. (1994) 'Colour and pattern in predator–prey interactions: the bright body colours and patterns of a tropical orb-spinning spider attract flower-seeking prey'. *Functional Ecology* 8: 616–20.
9. Tso, I.-M., Liao, C.-P., Huang, R.-P., and Yang, E.-C. (2006) 'Function of being colorful in web spiders: attracting prey or camouflaging oneself?' *Behavioral Ecology* 17: 606–13.
10. Fan, C.-M., Yang, E.-C., and Tso, I.-M. (2009) 'Hunting efficiency and predation risk shapes the color-associated foraging traits of a predator'. *Behavioral Ecology* 20: 808–16.
11. For example, work by Alex Bush and colleagues has shown that the coloration of wasp spiders (*Argiope bruennichi*) found in the UK serves to lure prey. Bush, A.A., Yu, D.W., and Herberstein, M.E. (2008) 'Function of bright coloration in the wasp spider *Argiope bruennichi* (Araneae: Araneidae)'. *Proceedings of the Royal Society of London B: Biological Sciences* 275: 1337–42.
12. Chuang, C.-Y., Yang, E.-C., and Tso, I.-M. (2007) 'Diurnal and nocturnal prey luring of a colorful predator'. *Journal of Experimental Biology* 210: 3830–7.
13. Tso, I.-M., Huang, J.-N., and Liao, C.-P. (2007) 'Nocturnal hunting of a brightly coloured sit-and-wait predator'. *Animal Behaviour* 74: 787–93.

14. This study was sophisticated in the way that it used digital image analysis to create images corresponding to the colours that might be seen by a pollinator such as a bee from varying distances. However, although the pattern information in the images was judged to qualitatively resemble the appearance of flowers (such as in providing symmetric patterns), no direct comparisons between spider and real flower shapes were made. Chiao, C.-C., Wu, W.-Y., Chen, S.-H., and Yang, E.-C. (2009) 'Visualization of the spatial and spectral signals of orb-weaving spiders, *Nephila pilipes*, through the eyes of the honeybee'. *Journal of Experimental Biology* 212: 2269–78.

15. Eberhard, W.G. (1977) 'Aggressive chemical mimicry by a bolas spider'. *Science* 198: 1173–5.

16. Stowe, M.K., Tumlinson, J.H., and Heath, R.R. (1987) 'Chemical mimicry: bolas spiders emit components of moth prey species pheromones'. *Science* 236: 964–6. This work was followed by other studies showing that the bolas spiders do not only mimic the presence of components found in moth pheromones, but they also match the relative ratios of the chemical components involved: Gemeno, C., Yeargan, K.V., and Haynes, K.F. (2000) 'Aggressive chemical mimicry by the bolas spider *Mastophora hutchinsoni*: identification and quantification of a major prey's sex pheromone components in the spider's volatile emissions'. *Journal of Chemical Ecology* 26: 1235–43.

17. Yeargan, K.V. (1988) 'Ecology of a bolas spider, *Mastophora hutchinsoni*: phenology, hunting tactics, and evidence for aggressive chemical mimicry'. *Oecologia* 74: 524–30; Haynes, K.F., Yeargan, K.V., and Gemeno, C. (2001) 'Detection of prey by a spider that aggressively mimics pheromone blends'. *Journal of Insect Behavior* 14: 535–44; and Haynes, K.F., Gemeno, C., Yeargan, K.V., Millar, J.G., and Johnson, K.M. (2002) 'Aggressive chemical mimicry of moth pheromones by a bolas spider: how does this specialist predator attract more than one species of prey?' *Chemoecology* 12: 99–105.

18. This was first investigated by James Lloyd in the 1960s. Lloyd, J.E. (1965) 'Aggressive mimicry in *Photuris*: firefly femmes fatales'. *Science* 149: 653–4.

19. Eisner, T., Goetz, M.A., Hill, D.E., et al. (1997) 'Firefly "femmes fatales" acquire defensive steroids (lucibufagins) from their firefly prey'. *Proceedings of the National Academy of Sciences of the USA* 94: 9723–8.

20. Jackson, R.R. and Blest, A.D. (1982) 'The biology of *Portia fimbriata*, a web-building jumping spider (Araneae: Salticidae) from Queensland: utilization of webs and predatory versatility'. *Journal of the Zoological Society of London* 196: 255–93; and Jackson, R.R. and Wilcox, R.S. (1990) 'Aggressive mimicry, prey-specific predatory behaviour and predator-recognition in the predator–prey interactions of *Portia fimbriata* and *Euryattus* sp., jumping spiders from Queensland'. *Behavioral Ecology and Sociobiology* 26: 111–19.

21. Tarsitano, M., Jackson, R.R., and Kirchner, W.H. (2000) 'Signals and signal choices made by the araneophagic jumping spider *Portia fimbriata* while hunting the orb-weaving web spiders *Zygiella x-notata* and *Zosis geniculatus*'. *Ethology* 106: 595–615.

22. De Serrano, A., Weadick, C., Price, A., and Rodd, F. (2012) 'Seeing orange: prawns tap into a pre-existing sensory bias of the Trinidadian guppy'. *Proceedings of the Royal Society B: Biological Sciences* 279: 3321–8.

23. Nelson, X.J., Garnett, D.T., and Evans, C.S. (2010) 'Receiver psychology and the design of the deceptive caudal luring signal of the death adder'. *Animal Behaviour* 79: 555–61.

24. Miya, M., Pietsch, T.W., and Orr, J.W. (2010) 'Evolutionary history of anglerfishes (Teleostei: Lophiiformes): a mitogenomic perspective'. *BMC Evolutionary Biology* 10: 58.

25. Pietsch, T.W. (1975) 'Precocious sexual parasitism in the deep sea ceratioid anglerfish, *Cryptopsaras couesi* Gill'. *Nature* 256: 38–40.

26. Haddock, S.H.D., Dunn, C.W., Pugh, P.R., and Schnitzler, C.E. (2005) 'Bioluminescent and red-fluorescent lures in a deep-sea siphonophore'. *Science* 309: 263.

27. Hoving, H.J.T., Zeidberg, L.D., Benfield, M.C., et al. (2013) 'First *in situ* observations of the deep-sea squid *Grimalditeuthis bonplandi* reveal unique use of tentacles'. *Proceedings of the Royal Society B: Biological Sciences* 280: 1463.

28. Darwin, C.R. (1875) *Insectivorous Plants*. London: John Murray.

29. Schaefer, H.M. and Ruxton, G.D. (2008) 'Fatal attraction: carnivorous plants roll out the red carpet to lure insects'. *Biology Letters* 4: 153–5.

30. Bennett, K.F. and Ellison, A.M. (2009) 'Nectar, not colour, may lure insects to their death'. *Biology Letters* 5: 469–72.

31. Kurup, R., Johnson, A.J., Sankar, S., Hussain, A.A., Santhish Kumar, C., et al. (2013) 'Fluorescent prey traps in carnivorous plants'. *Plant Biology* 15: 611–15.

32. Jürgens, A., El-Sayed, A.M., and Max Suckling, D. (2009) 'Do carnivorous plants use volatiles for attracting prey insects?' *Functional Ecology* 23: 875–87.

33. Kreuzwieser, J., Scheerer, U., Kruse, J., et al. (2014) 'The Venus flytrap attracts insects by the release of volatile organic compounds'. *Journal of Experimental Botany* 65: 755–66.

34. Di Giusto, B., Bessiere, J.-M., Gueroult, M., et al. (2010) 'Flower scent mimicry masks a deadly trap in the carnivorous plant *Nepenthes rafflesiana*'. *Journal of Ecology* 98: 845–56.

35. Bauer, U., Federle, W., Seidel, H., Grafe, U., and Ioannou, C. (2015) 'How to catch more prey with less effective traps: explaining the evolution of temporarily inactive traps in carnivorous pitcher plants'. *Proceedings of the Royal Society B: Biological Sciences* 282: 2675.

36. Schaefer, H.M. and Ruxton, G.D. (2009) 'Deception in plants: mimicry or perceptual exploitation?' *Trends in Ecology & Evolution* 24: 676–85.

CHAPTER 4

1. Dawkins, R. and Krebs, J. (1979) 'Arms races between and within species'. *Proceedings of the Royal Society of London B: Biological Sciences* 205: 489–511.

2. Stevens, M. (2013) *Sensory Ecology, Behaviour, and Evolution*. Oxford: Oxford University Press.

3. Schaefer, H.M. and Ruxton, G.D. (2011) *Plant–Animal Communication*. Oxford: Oxford University Press.

4. See for example: Behrens, R.R. (2002) *False Colors: Art, Design and Modern Camouflage*. Dysart, Iowa: Bobolink Books, and other works by the same author.

5. Wallace, A.R. (1889) *Darwinism: An Exposition of the Theory of Natural Selection With Some of its Applications*. London: Macmillan & Co.

6. Erasmus Darwin was by many measures a remarkable character: a physician by trade, part scientist (biology, chemistry, and physics), part poet and philosopher, and part engineer and inventor, becoming a Fellow of the Royal Society by the age of twenty-nine. Between 1794 and 1796 Erasmus published his vast two-volume biology and

medical book called *Zoonomia*. One of the most controversial aspects of this was his outline of biological evolution—a subject that repulsed many of the highly religious community at the time. He was highly regarded as a poet, and in his *The Temple of Nature* (1803) Erasmus sets out how life may have evolved and changed over time. Although his writings were descriptive (his grandson would provide the answer to evolution's mechanisms), he clearly understood much about adaptation, including camouflage. See: King-Hele, D.G. (1988) 'Erasmus Darwin, man of ideas and inventor of words'. *Notes and Records of the Royal Society of London* 42: 149–80.

7. A detailed obituary of Poulton's life and achievements can be found here: Carpenter, G.D.H. (1944) 'Edward Bagnall Poulton: 1856–1943'. *Obituary Notices of Fellows of the Royal Society* 4: 655–80.

8. Poulton, E.B. (1890) *The Colours of Animals: Their Meaning and Use. Especially Considered in the Case of Insects*. Second Edition. London: Kegan Paul, Trench Trübner, & Co. Ltd.

9. A detailed account of the peppered moth and the research that investigated it up until the 1990s can be found in: Majerus, M.E.N. (1998) *Melanism: Evolution in Action*. Oxford: Oxford University Press. Another somewhat more recent review is: Cook, L.M. (2003) 'The rise and fall of the carbonaria form of the peppered moth'. *The Quarterly Review of Biology* 78: 399–417.

10. Grant, B.S., Owen, D.F., and Clark, C.A. (1996) 'Parallel rise and fall of melanic peppered moths in America and Britain'. *Journal of Heredity* 87: 351–7.

11. Kettlewell published a number of studies on the peppered moth, but the seminal work included: Kettlewell, H.B.D. (1955) 'Selection experiments on industrial melanism in the Lepidoptera'. *Heredity* 9: 323–42; and Kettlewell, H.B.D. (1956) 'Further selection experiments on industrial melanism in the Lepidoptera'. *Heredity* 10: 287–301.

12. The popularization of Kettlewell's and Ford's work on the peppered moth, specifically to present the idea that they committed fraud, was published in the early 2000s: Hooper, J. (2002) *Of Moths and Men: Intrigue, Tragedy and the Peppered Moth*. London: Fourth Estate. The story was rapidly latched on to by creationists at the time and ever since as a route to attack evolution and its teaching. However, numerous reviews of the book comprehensively showed that the claims of fraud and the majority of criticisms regarding Kettlewell's work are entirely without foundation, and pointed out a range of errors and misinterpretations by Hooper. They also reinforced the fact that there is no reason to doubt the overall conclusions of Kettlewell's work, in that changes in the frequency of peppered moth morphs was due to camouflage and differential predation by birds. For example, see: Coyne, J.A. (2002) 'Evolution under pressure: a look at the controversy about industrial melanism in the peppered moth. Review of Hooper 2002, *Of Moths and Men: Intrigue, Tragedy and the Peppered Moth*'. *Nature* 418: 19–20; Grant, B.S. (2002) 'Sour grapes of wrath. Review of Hooper 2002, *Of Moths and Men: Intrigue, Tragedy and the Peppered Moth*'. *Science* 297: 940–1; Rudge, D.W. (2005) 'Did Kettlewell commit fraud? Re-examining the evidence'. *Public Understanding of Science* 14: 249–68; and Rudge, D.W. (2006) 'Myths about moths: a study in contrasts'. *Endeavour* 30: 19–23.

13. Majerus, M.E.N., Brunton, C.F.A., and Stalker, J. (2000) 'A bird's eye view of the peppered moth'. *Journal of Evolutionary Biology* 13: 155–9.

14. Cook, L.M., Mani, G.S., and Varley, M.E. (1986) 'Postindustrial melanism in the peppered moth'. *Science* 231: 611–13.

15. By a stroke of good fortune, Majerus presented his studies, including the results and data, at an evolutionary biology conference in Sweden in 2007, and he made his slides freely available on the Internet thereafter. He also published his methods in an earlier book chapter. The work was eventually published as: Cook, L.M., Grant, B.S., Saccheri, I.J., and Mallet, J. (2012) 'Selective bird predation on the peppered moth: the last experiment of Michael Majerus'. *Biology Letters* 8: 609–12.

16. Beyond the survival data, Majerus' study also included 135 observations of natural resting sites of wild moths (different to those used in his experiment). This showed that, contrary to some criticism that the moths might not rest on tree trunks as Kettlewell supposed, a considerable number did after all. In addition, there was no difference in chosen resting sites between the different forms of the peppered moth.

17. Wallace, A.R. (1867) 'Mimicry and other protective resemblances among animals'. *Westminster Review* (London ed.) 1 July: 1–43.

18. For example, see: Nachman, M.W., Hoekstra, H.E., and D'Agostino, S.L. (2003) 'The genetic basis of adaptive melanism in pocket mice'. *Proceedings of the National Academy of Sciences of the USA* 100: 5268–73; and Manceau, M., Domingues, V.S., Mallarino, R., and Hoekstra, H.E. (2011) 'The developmental role of agouti in color pattern evolution'. *Science* 331: 1062–5.

19. For example, see: Rosenblum, E.B., Römpler, H., Schöneberg, T., and Hoekstra, H.E. (2010) 'Molecular and functional basis of phenotypic convergence in white lizards at White Sands'. *Proceedings of the National Academy of Sciences of the USA* 107: 2113–17.

20. Vignieri, S.N., Larson, J.G., and Hoekstra, H.E. (2010) 'The selective advantage of crypsis in mice'. *Evolution* 64: 2153–8.

21. Todd, P.A., Briers, R.A., Ladle, R.J., and Middleton, F. (2006) 'Phenotype-environment matching in the shore crab (*Carcinus maenas*)'. *Marine Biology* 148: 1357–67; and Stevens, M., Lown, A.E., and Wood, L.E. (2014) 'Camouflage and individual variation in shore crabs (*Carcinus maenas*) from different habitats'. *PLoS ONE* 9: e115586.

22. Pietrewicz, A.T. and Kamil, A.C. (1977) 'Visual detection of cryptic prey by blue jays (*Cyanocitta cristata*)'. *Science* 195: 580–2.

23. Pietrewicz, A.T. and Kamil, A.C. (1979) 'Search image formation in the blue jay (*Cyanocitta cristata*)'. *Science* 204: 1332–3.

24. Bond, A.B. and Kamil, A.C. (1998) 'Apostatic selection by blue jays produces balanced polymorphism in virtual prey'. *Nature* 395: 594–6; Bond, A.B. and Kamil, A.C. (2002) 'Visual predators select for crypticity and polymorphism in virtual prey'. *Nature* 415: 609–13; and Bond, A.B. and Kamil, A.C. (2006) 'Spatial heterogeneity, predator cognition, and the evolution of color polymorphism in virtual prey'. *Proceedings of the National Academy of Sciences of the USA* 103: 3214–19.

25. Kettlewell, H.B.D. (1955) 'Recognition of appropriate backgrounds by the pale and black phases of Lepidoptera'. *Nature* 175: 943–4. Kettlewell also published similar experiments around twenty years later with several other species, with similar findings: Kettlewell, H.B.D. and Conn, D.L.T. (1977) 'Further background-choice experiments on cryptic Lepidoptera'. *Journal of Zoology* 181: 371–6.

26. Sargent, T.D. (1966) 'Background selections of geometrid and noctuid moths'. *Science* 154: 1674–5.

27. Sargent, T.D. (1968) 'Cryptic moths: effects on background selections of painting the circumocular scales'. *Science* 159: 100–1.

28. Lovell, P.G., Ruxton, G.D., Langridge, K.V., and Spencer, K.A. (2013) 'Individual quail select egg-laying substrate providing optimal camouflage for their egg phenotype'. *Current Biology* 23: 260–4.

29. Sargent, T.D. (1969) 'Behavioural adaptations of cryptic moths III: resting attitudes of two bark-like species, *Melanolophia canadaria* and *Catocala ultronia*'. *Animal Behaviour* 17: 670–2.

30. Kang, C.K., Moon, J.Y., Lee, S.I., and Jablonski, P.G. (2012) 'Camouflage through an active choice of a resting spot and body orientation in moths'. *Journal of Evolutionary Biology* 25: 1695–702; Kang, C.K., Moon, J.Y., Lee, S.I., and Jablonski, P.G. (2013) 'Moths on tree trunks seek out more cryptic positions when their current crypticity is low'. Animal Behaviour 86: 587–94; and Kang, C.K., Stevens, M., Moon, J.Y., Lee, S.I., and Jablonski, P.G. (2015) 'Camouflage through behavior in moths: the role of background matching and disruptive coloration'. *Behavioral Ecology* 26: 45–54.

31. For a detailed review of their colour change for camouflage, see: Hanlon, R.T., Chiao, C.-C., Mäthger, L.M., Barbosa, A., Buresch, K.C., and Chubb C. 2009. 'Cephalopod dynamic camouflage: bridging the continuum between background matching and disruptive coloration'. *Philosophical Transactions of the Royal Society B: Biological Sciences* 364: 429–37.

32. See for example, Barbosa, A., Mäthger, L.M., Buresch, K.C., Kelly, J., Chubb, C., et al. (2008) 'Cuttlefish camouflage: the effects of substrate contrast and size in evoking uniform, mottle or disruptive body patterns'. *Vision Research* 48: 1242–53, and Chiao, C.C., Wickiser, J.K., Allen, J.J., Genter, B., and Hanlon, R.T. (2011) 'Hyperspectral imaging of cuttlefish camouflage indicates good color match in the eyes of fish predators'. *Proceedings of the National Academy of Sciences of the USA* 108: 9148–53.

33. Wallace, A.R. (1877) 'The colours of animals and plants'. *Macmillan's Magazine* September and October: 384–471.

34. Poulton's work on caterpillars went some way to showing that diet, and perhaps especially background colour, was important in determining the colour morph of caterpillars from several species. For example, see: Poulton, E.B. (1885) 'The essential nature of the colouring of phytophagous larvae (and their pupae); with an account of some experiments upon the relation between the colour of such larvae and that of their food-plants'. *Proceedings of the Royal Society B: Biological Sciences* 237: 269–315; Poulton, E.B. (1887) 'An enquiry into the cause and extent of a special colour relation between certain exposed lepidopterous pupae and the surfaces which immediately surround them'. *Philosophical Transactions of the Royal Society B: Biological Sciences* 178: 311–441; and Poulton, E.B. (1903) 'Experiments in 1893, 1894, and 1896 upon the colour-relation between lepidopterous larvae and their surroundings, and especially the effect of lichen-covered bark upon *Odontopera bidentata*, *Gastropacha quercifolia*, etc.' *Transactions of the Entomological Society of London* 1903: 311–74. More recent work has also tested the effects of background on diet, finding that both can be important, depending on the species. For example: Greene, E. (1996) 'Effect of light quality and larval diet on morph

induction in the polymorphic caterpillar *Nemoria arizonaria* (Lepidoptera: Geometridae)'. *Biological Journal of the Linnean Society* 58: 277–85; and Noor, M.A.F., Parnell, R.S., and Grant, B.S. (2008) 'A reversible color polyphenism in American peppered moth (*Biston betularia cognataria*) caterpillars'. *PLoS ONE* 3: e3142.

35. Stevens, M., Lown, A.E., and Wood, L.E. (2014) 'Colour change and camouflage in juvenile shore crabs *Carcinus maenas*'. *Frontiers in Ecology and Evolution* 2: 14.

36. Stevens, M., Lown, A.E., and Denton, A.M. (2014) 'Rockpool gobies change colour for camouflage'. *PLoS ONE* 9: e110325.

37. Thayer, G.H. with Thayer, A.H. (1909) *Concealing-Coloration in the Animal Kingdom: An Exposition of the Laws of Disguise Through Color and Pattern: Being a Summary of Abbott H. Thayer's Discoveries*. New York: Macmillan.

38. Much information regarding the role of Thayer and others in attempting to influence military camouflage can be found for example in: Forbes, P. (2009) *Dazzled and Deceived: Mimicry and Camouflage*. New Haven, CT: Yale University Press; and Behrens, R.R. (2009) 'Revisiting Abbott Thayer: non-scientific reflections about camouflage in art, war and zoology'. *Philosophical Transactions of the Royal Society, Series B* 364: 497–501.

39. Gould, S.J. (1991) 'Red Wings in the Sunset'. In *Bully for Brontosaurus*. London: Penguin Books.

40. The debates between Thayer, Roosevelt, and others have been covered in some detail in several other publications, including: Kingsland, S. (1978) 'Abbott Thayer and the protective coloration debate'. *Journal of the History of Biology* 11: 223–44; and Nemerov, A. (1997) 'Vanishing Americans: Abbott Thayer, Theodore Roosevelt, and the attraction of camouflage'. *American Art* 11: 50–81.

41. Cott, H.B. (1940) *Adaptive Coloration in Animals*. London: Methuen & Co.

42. See for example: Forsyth, I. (2014) 'The practice and poetics of fieldwork: Hugh Cott and the study of camouflage'. *Journal of Historical Geography* 43: 128–37.

43. For example: Cuthill, I.C., Stevens, M., Sheppard, J., Maddocks, T., Párraga, C.A., et al. (2005) 'Disruptive coloration and background pattern matching'. *Nature* 434: 72–4; and Stevens, M., Cuthill, I.C., Windsor, A.M.M., and Walker, H.J. (2006) 'Disruptive contrast in animal camouflage'. *Proceedings of the Royal Society of London B: Biological Sciences* 273: 2433–8.

44. Stevens, M. and Cuthill, I.C. (2006) 'Disruptive coloration, crypsis and edge detection in early visual processing'. *Proceedings of the Royal Society B: Biological Sciences* 273: 2141–7.

45. Webster, R.J., Hassall, C., Herdman, C.M., and Sherratt, T.N. (2013) 'Disruptive camouflage impairs object recognition'. *Biology Letters* 9: 0501.

46. One other type of camouflage involves countershading. Here, the surface of an animal facing the highest light intensity from the sun (normally the top surface) is darker than the surfaces facing away from the light. This pattern is found across many animal groups, both aquatic and invertebrate. Poulton and especially Thayer were the first to argue that countershading worked in concealment to cancel out the effect of shadows created by an animal's own body and to hide three-dimensional shape cues that could be used in detection. Like disruptive coloration, it was only recently that clear evidence that countershading works in camouflage has been found, including work by Hannah Rowland, who at the time was a PhD student at the University of Liverpool: Rowland,

H.M., Speed, M.P., Ruxton, G.D., Edmunds, M., Stevens, M., et al. (2007) 'Countershading enhances cryptic protection: an experiment with wild birds and artificial prey'. *Animal Behaviour* 74: 1249–58; and Rowland, H.M., Cuthill, I.C., Harvey, I.F., Speed, M.P., and Ruxton, G.D. (2008) 'Can't tell the caterpillars from the trees: countershading enhances survival in a woodland'. *Proceedings of the Royal Society of London B: Biological Sciences* 275: 2539–45.

47. Beddard, F.E. (1892) *Animal Coloration; An Account of the Principle Facts and Theories Relating to the Colours and Markings of Animals.* London: Swan Sonnenschein.

48. See: Skelhorn, J., Rowland, H.M., Speed, M.P., and Ruxton, G.D. (2010) 'Masquerade: camouflage without crypsis'. *Science* 327: 51; Skelhorn, J. and Ruxton, G.D. (2010) 'Predators are less likely to misclassify masquerading prey when their models are present'. *Biology Letters* 6: 597–9; and Skelhorn, J., Rowland, H.M., Delf, J., Speed, M.P., and Ruxton, G.D. (2011) 'Density-dependent predation influences the evolution and behavior of masquerading prey'. *Proceedings of the National Academy of Sciences of the USA* 108: 6532–6.

49. Liu, M.-H., Blamires, S.J., Liao, C.-P., and Tso, I.-M. (2014) 'Evidence of bird dropping masquerading by a spider to avoid predators'. *Scientific Reports* 4: 5058.

50. Klomp, D.A., Stuart-Fox, D., Das, I., and Ord, T.J. (2014) 'Marked colour divergence in the gliding membranes of a tropical lizard mirrors population differences in the colour of falling leaves'. *Biology Letters* 10: 0776.

51. Many others beyond Thayer were involved in designing ship camouflage, including other artists. See: Behrens, R.R. (1999) 'The role of artists in ship camouflage during World War I'. *Leonardo* 32: 53–9.

52. Stevens, M., Yule, D.H., and Ruxton, G.D. (2008) 'Dazzle coloration and prey movement'. *Proceedings of the Royal Society of London B: Biological Sciences* 275: 2639–43; and Stevens, M., Searle, W.T.L., Seymour, J.E., Marshall, K.L.A., and Ruxton, G.D. (2011) 'Motion dazzle and camouflage as distinct anti-predator defenses'. *BMC Biology* 9: 81.

53. Scott-Samuel, N.E., Baddeley, R., Palmer, C.E., and Cuthill, I.C. (2011) 'Dazzle camouflage affects speed perception'. *PLoS ONE* 6: e20233.

54. How, M.J. and Zanker, J.M. (2014) 'Motion camouflage induced by zebra stripes'. *Zoology* 117: 163–70.

55. Caro, T., Izzo, A., Reiner, R.C., Walker, H., and Stankowich, T. (2014) 'The function of zebra stripes'. *Nature Communications* 5: 3535.

56. Behrens, R.R. (2009) 'Revisiting Abbott Thayer: non-scientific reflections about camouflage in art, war and zoology'. *Philosophical Transactions of the Royal Society, Series B* 364: 497–501.

57. Several studies have now demonstrated that the concept of camouflage can apply to hiding the odour of animals too, by resembling the smell of the background. In some cases, including some fish, this can work based on the diet of the animal: Brooker, R.M., Munday, P.L., Chivers, D.P., and Jones, G.P. (2015) 'You are what you eat: diet-induced chemical crypsis in a coral-feeding reef fish'. *Proceedings of the Royal Society B: Biological Sciences* 282: 1887.

58. Recent work has shown that some plants appear to be camouflaged against the background. For example: Niu, Y., Chen, G., Peng, D.-L., Song, B., Yang, Y., et al.

(2014) 'Grey leaves in an alpine plant: a cryptic colouration to avoid attack?' *New Phytologist* 203: 953–63. Plant camouflage can include the seeds, which in some species of plant have different colours depending on the appearance of the soil: Porter, S.S. (2013) 'Adaptive divergence in seed color camouflage in contrasting soil environments'. *New Phytologist* 197: 1311–20.

CHAPTER 5

1. Darwin, C.R. (1892). In F. Darwin, editor. *Selected Letters on Evolution and Origin of Species, with an Autobiographical Chapter.* Mineola, NY: Dover Publications Inc.

2. Wallace set out his ideas of warning coloration in several places, including back in 1867 and more fully in 1877 (and subsequently thereafter): Wallace, A.R. (1867) 'Mimicry and other protective resemblances among animals'. *Westminster Review* (London ed.) 1 July: 1–43; Wallace, A.R. (1877) 'The colours of animals and plants'. *Macmillan's Magazine* September and October: 384–471.

3. Poulton, E.B. (1890) *The Colours of Animals: Their Meaning and Use. Especially Considered in the Case of Insects.* Second Edition. London: Kegan Paul, Trench Trübner, & Co. Ltd.

4. Much information regarding warning signals and how they work can be found in: Ruxton, G.D., Sherratt, T.N., and Speed, M.P. (2004) *Avoiding Attack.* Oxford: Oxford University Press, and numerous other sources.

5. Bates first presented his paper to the Linnean Society in 1861, and then published his findings and ideas in 1862: Bates, H.W. (1862) 'Contributions to an insect fauna of the Amazon valley. Lepidoptera: Heliconidae'. *Transactions of the Linnean Society of London* 23: 495–566.

6. C.R. Darwin to H.W. Bates, 20 November 1862.

7. Mostler, G. (1935) 'Observations on the question of wasp mimicry'. *Z Morph Okol Tiere* 29: 381–454.

8. Golding, Y., Ennos, R., Sullivan, M., and Edmunds, M. (2015) 'Hoverfly mimicry deceives humans'. *Journal of Zoology* 266: 395–9.

9. Dittrich, W., Gilbert, F., Green, P., McGregor, P., and Grewcock, D. (1993) 'Imperfect mimicry: a pigeon's perspective'. *Proceedings of the Royal Society of London B: Biological Sciences* 251: 195–200.

10. See for example: Cuthill, I.C. (2006) 'Color perception'. In: G.E. Hill and K.J. McGraw, editors. *Bird Coloration, Volume 1: Methods and Mechanisms.* Cambridge, MA: Harvard University Press.

11. Bain, R.S., Rashed, A., Cowper, V.J., Gilbert, F.S., and Sherratt, T.N. (2007) 'The key mimetic features of hoverflies through avian eyes'. *Proceedings of the Royal Society B: Biological Sciences* 274: 1949–54.

12. Rashed, A., Khan, M.I., Dawson, J.W., Yack, J.E., and Sherratt, T.N. (2009) 'Do hoverflies (Diptera: Syrphidae) sound like the Hymenoptera they morphologically resemble?' *Behavioral Ecology* 20: 396–402.

13. Penney, H.D., Hassall, C., Skevington, J.H., Lamborn, B., and Sherratt, T.N. (2014) 'The relationship between morphological and behavioral mimicry in hover flies (Diptera: Syrphidae)'. *American Naturalist* 183: 281–9.

14. Edmunds, M. and Reader, T. (2014) 'Evidence for Batesian mimicry in a polymorphic hoverfly'. *Evolution* 68: 827–39.

15. Howarth, B., Edmunds, M., and Gilbert, F. (2004) 'Does the abundance of hoverfly (Syrphidae) mimics depend on the numbers of their hymenopteran models?' *Evolution* 58: 367–75.

16. Wallace, A.R. (1867) 'Mimicry and other protective resemblances among animals'. *Westminster Review* (London ed.) 1 July: 1–43. Here Wallace also references Bates as having discovered a mantis that mimics ants in the Amazon.

17. Nelson, X.J. (2012) 'A predator's perspective of the accuracy of ant mimicry in spiders'. *Psyche* 2012: 1–5.

18. Nelson, X.J. and Jackson, R.R. (2006) 'Vision-based innate aversion to ants and ant mimics'. *Behavioral Ecology* 17: 676–81.

19. Nelson, X.J. and Jackson, R.R. (2006) 'Compound mimicry and trading predators by the males of sexually dimorphic Batesian mimics'. *Proceedings of the Royal Society of London B: Biological Sciences* 273: 367–72.

20. Nelson, X.J. and Jackson, R.R. (2009) 'Collective Batesian mimicry of ant groups by aggregating spiders'. *Animal Behaviour* 78: 123–9.

21. Rowe, M.P., Coss, R.G., and Owings, D.H. (1986) 'Rattlesnake rattles and burrowing owl hisses: a case of acoustic Batesian mimicry'. *Ethology* 72: 53–71.

22. Barber, J.R. and Conner, W.E. (2007) 'Acoustic mimicry in a predator–prey interaction'. *Proceedings of the National Academy of Sciences of the USA* 104: 9331–4.

23. Corcoran, A.J. and Hristov, N.I. (2014) 'Convergent evolution of anti-bat sounds'. *Journal of Comparative Physiology A* 200: 811–21.

24. Gilbert, F. (2005) 'The evolution of imperfect mimicry'. In M. Fellowes, G. Holloway, and J. Rolff, editors. *Insect Evolutionary Ecology*. Wallingford: CABI.

25. Cuthill, I.C. and Bennett, A.T.D. (1993) 'Mimicry and the eye of the beholder'. *Proceedings of the Royal Society of London B: Biological Sciences* 253: 203–4.

26. Green, P.R., Gentle, L., Peake, T.M., Scudamore, R.E., McGregor, P., et al. (1999) 'Conditioning pigeons to discriminate naturally lit insect specimens'. *Behavioural Processes* 46: 97–102.

27. Kazemi, B., Gamberale-Stille, G., Tullberg, B.S., and Leimar, O. (2014) 'Stimulus salience as an explanation for imperfect mimicry'. *Current Biology* 24: 965–9.

28. Kikuchi, D.W. and Pfennig, D.W. (2010) 'Predator cognition permits imperfect coral snake mimicry'. *American Naturalist* 176: 830–4.

29. Harper, G.R. and Pfennig, D.W. (2007) 'Mimicry on the edge: why do mimics vary in resemblance to their model in different parts of their geographical range?' *Proceedings of the Royal Society B: Biological Sciences* 274: 1955–61.

30. Penney, H.D., Hassall, C., Skevington, J.H., Abbott, K.R., and Sherratt, T.N. (2012) 'A comparative analysis of the evolution of imperfect mimicry'. *Nature* 483: 461–4.

31. Londono, G.A., García, D.A., and Sánchez Martínez, M.A. (2015) 'Morphological and behavioral evidence of Batesian mimicry in nestlings of a lowland Amazonian bird'. *American Naturalist* 185: 135–41.

32. Huey, R.B. and Pianku, E.R. (1977) 'Natural selection for juvenile lizards mimicking noxious beetles'. *Science* 193: 201–2.

CHAPTER 6

1. Humphries, D.A. and Driver, P.M. (1967) 'Erratic display as a device against predators'. *Science* 156: 1767–8; and Humphries, D.A. and Driver, P.M. (1970) 'Protean defence by prey animals'. *Oecologia* 5: 285–302.
2. Sargent, T.D. (1969) 'A suggestion regarding the hindwing diversity among moths of the genus *Catocala* (Noctuidae)'. *Journal of the Lepidopterists' Society* 23: 261–4; Sargent, T.D. (1976) *Legion of Night: the Underwing Moths*. Amherst: University of Massachusetts Press; and Sargent, T.D. (1978) 'On the maintenance of stability in hindwing diversity among moths of the genus *Catocala* (Lepidoptera: Noctuidae)'. *Evolution* 424–34.
3. Vaughan, F.A. (1983) 'Startle responses of blue jays to visual stimuli presented during feeding'. *Animal Behaviour* 31: 385–96.
4. Schlenoff, D.H. (1985) 'The startle responses of blue jays to *Catocala* (Lepidoptera: Noctuidae) prey models'. *Animal Behaviour* 33: 1057–67.
5. Ingalls, V. (1993) 'Startle and habituation responses of blue jays (*Cyanocitta cristata*) in a laboratory simulation of anti-predator defenses of *Catocala* moths (Lepidoptera: Noctuidae)'. *Behaviour* 126: 77–96.
6. Langridge, K. (2009) 'Cuttlefish use startle displays, but not against large predators'. *Animal Behaviour* 77: 847–56; and Langridge, K., Broom, M., and Osorio, D. (2007) 'Selective signalling by cuttlefish to predators'. *Current Biology* 17: R1044–R1045.
7. Bates, D.L. and Fenton, M.B. (1990) 'Aposematism or startle? Predators learn their responses to the defenses of prey'. *Canadian Journal of Zoology* 68: 49–52.
8. Yager, D.D. and Spangler, H.G. (1997) 'Behavioral response to ultrasound by the tiger beetle *Cicindela marutha* Dow combines aerodynamic changes and sound production'. *Journal of Experimental Biology* 200: 649–59.
9. Bura, V.L., Rohwer, V.G., Martin, P.R., and Yack, J.E. (2010) 'Whistling in caterpillars (*Amorpha juglandis*, Bombycoidea): sound-producing mechanism and function'. *Journal of Experimental Biology* 214: 30–7.
10. Olofsson, M., Jakobsson, S., and Wiklund, C. (2012) 'Auditory defence in the peacock butterfly (*Inachis io*) against mice (*Apodemus flavicollis* and *A. sylvaticus*)'. *Behavioral Ecology and Sociobiology* 66: 209–15.
11. Blest, A.D. (1957) 'The function of eyespot patterns in the Lepidoptera'. *Behaviour* 11: 209–56.
12. Vallin, A., Jakobsson, S., Lind, J., and Wiklund, C. (2005) 'Prey survival by predator intimidation: an experimental study of peacock butterfly defence against blue tits'. *Proceedings of the Royal Society of London B: Biological Sciences* 272: 1203–7.
13. Vallin, A., Jakobsson, S., and Wiklund, C. (2007) '"An eye for an eye?"—on the generality of the intimidating quality of eyespots in a butterfly and a hawkmoth'. *Behavioral Ecology and Sociobiology* 61: 1419–24.
14. Kirby, W. and Spence, W. (1818) *An Introduction to Entomology, or Elements of the Natural History of Insects*. With Plates. London: Longman.
15. Stevens, M., Hopkins, E., Hinde, W., Adcock, A., Connelly, Y., et al. (2007) 'Field experiments on the effectiveness of "eyespots" as predator deterrents'. *Animal Behaviour* 74: 1215–27; Stevens, M., Hardman, C.J., and Stubbins, C.L. (2008) 'Conspicuousness,

not eye mimicry, makes "eyespots" effective anti-predator signals'. *Behavioral Ecology* 19: 525–31; and Stevens, M., Cantor, A., Graham, J., and Winney, I.S. (2009) 'The function of animal "eyespots": conspicuousness but not eye mimicry is key'. *Current Zoology* 55: 319–26.

16. See for example: Stevens, M. (2005) 'The role of eyespots as anti-predator mechanisms, principally demonstrated in the Lepidoptera.' *Biological Reviews* 80: 573–88; and Stevens, M. and Ruxton, G.D. (2014) 'Do animal eyespots really mimic eyes?' *Current Zoology* 60: 26–36.

17. A handful of studies I have not discussed have also aimed to test whether eyespots mimic eyes, including: Merilaita, S., Vallin, A., Kodandaramaiah, U., Dimitrova, M., Ruuskanen, S., et al. (2011) 'Number of eyespots and their intimidating effect on naive predators in the peacock butterfly'. *Behavioral Ecology* 22: 1326–31; and Olofsson, M., Løvlie, H., Tibblin, J., Jakobsson, S., and Wiklund, C. (2013) 'Eyespot display in the peacock butterfly triggers antipredator behaviours in naive adult fowl'. *Behavioral Ecology* 24: 305–10. However, these have generally not distinguished between the competing theories of eye mimicry and conspicuousness, see: Stevens, M. and Ruxton, G.D. (2014) 'Do animal eyespots really mimic eyes?' *Current Zoology* 60: 26–36.

18. Mukherjee, R. and Kodandaramaiah, U. (2015) 'What makes eyespots intimidating—the importance of pairedness'. *BMC Evolutionary Biology* 15: 34.

19. Blut, C., Wilbrandt, J., Fels, D., Girgel, E.I., and Lunau, K. (2012) 'The "sparkle" in fake eyes—the protective effect of mimic eyespots in lepidoptera'. *Entomologia* 143: 231–44.

20. De Bona, S., Valkonen, J.K., López-Sepulcre, A., and Mappes, J. (2015) 'Predator mimicry, not conspicuousness, explains the efficacy of butterfly mimicry'. *Proceedings of the Royal Society B: Biological Sciences* 282: 0202.

21. Bates, H.W. (1862) 'Contributions to an insect fauna of the Amazon valley (Lepidoptera: Heliconidae)'. *Transactions of the Linnean Society of London* 23: 495–566.

22. Hossie, T.J. and Sherratt, T.N. (2012) 'Eyespots interact with body colour to protect caterpillar-like prey from avian predators'. *Animal Behaviour* 84: 167–73; and Hossie, T.J. and Sherratt, T.N. (2013) 'Defensive posture and eyespots deter avian predators from attacking caterpillar models'. *Animal Behaviour* 86: 383–9.

23. Skelhorn, J., Dorrington, G., Hossie, T.J., and Sherratt, T.N. (2014) 'The position of eyespots and thickened segments influence their protective value to caterpillars'. *Behavioral Ecology* 25: 1417–22.

24. For example, see: Lyytinen, A., Brakefield, P.M., and Mappes, J. (2003) 'Significance of butterfly eyespots as an antipredator device in ground-based and aerial attacks'. *Oikos* 100: 373–9; Lyytinen, A., Brakefield, P.M., Lindström, L., and Mappes, J. (2004) 'Does predation maintain eyespot plasticity in *Bicyclus anynana*?' *Proceedings of the Royal Society of London B: Biological Sciences* 271: 279–83; and Vlieger, L. and Brakefield, P.M. (2007) 'The deflection hypothesis: eyespots on the margins of butterfly wings do not influence predation by lizards'. *Biological Journal of the Linnean Society* 92: 661–7.

25. Olofsson, M., Vallin, A., Jakobsson, S., and Wiklund, C. (2010) 'Marginal eyespots on butterfly wings deflect bird attacks under low light intensities with UV wavelengths'. *PLoS ONE* 5: e10798.

26. Olofsson, M., Jakobsson, S., and Wiklund, C. (2013) 'Bird attacks on a butterfly with marginal eyespots and the role of prey concealment against the background'. *Biological Journal of the Linnean Society* 109: 290–7.

27. Prudic, K.L., Stoehr, A.M., Wasik, B.R., and Monteiro, A. (2015) 'Eyespots deflect predator attack increasing fitness and promoting the evolution of phenotypic plasticity'. *Proceedings of the Royal Society B: Biological Sciences* 282: 1531.

28. Although some interesting studies have investigated this idea, they have generally not conducted predation experiments with butterflies with false heads, for example: Robbins, R.K. (1980) 'The lycaenid "false head" hypothesis: historical review and quantitative analysis'. *Journal of the Lepidopterists Society* 34: 194–208; and Wourms, M.K. and Wasserman, F.E. (1985) 'Butterfly wing markings are more advantageous during handling than during the initial strike of an avian predator'. *Evolution* 39: 845–51.

29. Van Buskirk, J., Aschwanden, J., Buckelmüller, I., Reolon, S., and Rüttiman, S. (2004) 'Bold tail coloration protects tadpoles from dragonfly strikes'. *Copeia* 3: 599–602.

30. Lönnstedt, O.M., McCormic, M.I., and Chivers, D.P. (2013) 'Predator-induced changes in the growth of eyes and false eyespots'. *Scientific Reports* 3: 2259.

31. Barber, J.R., Leavell, B.C., Keener, A.L., Breinholt, J.W., Chadwell, B.A., et al. (2015) 'Moth tails divert bat attack: evolution of acoustic deflection'. *Proceedings of the National Academy of Sciences of the USA* 112: 2812–16.

32. See: Miller, L.A. (1991) 'Arctiid moth clicks can degrade the accuracy of range difference discrimination in echolocating big brown bats, *Eptesicus fuscus*'. *Journal of Comparative Physiology A* 168: 571–9; Fullard, J.H., Simmons, J.A., and Saillant, P.A. (1994) 'Jamming bat echolocation: the dogbane tiger moth *Cycnia tenera* times its clicks to the terminal attack calls of the big brown bat *Eptesicus fuscus*'. *Journal of Experimental Biology* 194: 285–98; and Tougaard, J., Casseday, J.H., and Covey, E. (1998) 'Arctiid moths and bat echolocation: broad-band clicks interfere with neural responses to auditory stimuli in the nuclei of the lateral lemniscus of the big brown bat'. *Journal of Comparative Physiology A* 182: 203–15.

33. Corcoran, A.J., Barber, J.R., and Conner, W.E. (2009) 'Tiger moth jams bat sonar'. *Science* 325: 325–7; and Corcoran, A.J., Barber, J.R., Hristov, N.I., and Conner, W.E. (2011) 'How do tiger moths jam bat sonar?' *Journal of Experimental Biology* 214: 2416–25.

34. Further information about these types of behaviours can be found in: Ruxton, G.D., Sherratt, T.N., and Speed, M.P. (2004) *Avoiding Attack*. Oxford: Oxford University Press; and Caro, T. (2005) *Antipredator Defenses in Birds and Mammals*. Chicago: Chicago University Press.

CHAPTER 7

1. Tanaka, K.D. and Ueda, K. (2005) 'Horsfield's hawk-cuckoo nestlings simulate multiple gapes for begging'. *Science* 308: 653; and Tanaka, K.D., Morimoto, G., and Ueda, K. (2005) 'Yellow wing-patch of a nestling Horsfield's hawk cuckoo *Cuculus fugax* induces miscognition by hosts: mimicking a gape?' *Journal of Avian Biology* 36: 461–4.

2. See: Davies, N.B. (2000) *Cuckoos, Cowbirds and Other Cheats*. London: T. & A. D. Poyser; Davies, N.B. (2011) 'Cuckoo adaptations: trickery and tuning'. *Journal of Zoology* 284: 1–14; and Kilner, R.M. and Langmore, N.E. (2011) 'Cuckoos versus hosts in insects and birds: adaptations, counter-adaptations and outcomes'. *Biological Reviews* 86: 836–52.

3. Krüger, O., Sorenson, M.D., and Davies, N.B. (2009) 'Does coevolution promote species richness in parasitic cuckoos?' *Proceedings of the Royal Society B: Biological Sciences* 276: 3871–9.

4. Welbergen, J.A. and Davies, N.B. (2009) 'Strategic variation in mobbing as a front line of defense against brood parasitism'. *Current Biology* 19: 235–40.

5. Welbergen, J.A. and Davies, N.B. (2008) 'Reed warblers discriminate cuckoos from sparrowhawks with graded alarm signals that attract mates and neighbours'. *Animal Behaviour* 76: 811–22.

6. Wallace, A.R. (1867) 'Mimicry and other protective resemblances among animals'. *Westminster Review* (London ed.) 1 July: 1–43; and Wallace, A.R. (1889) *Darwinism: An Exposition of the Theory of Natural Selection With Some of its Applications*. London: Macmillan & Co.

7. Davies, N.B. and Welbergen, J.A. (2008) 'Cuckoo-hawk mimicry? An experimental test'. *Proceedings of the Royal Society of London B: Biological Sciences* 275: 1817–22; and Welbergen, J.A. and Davies, N.B. (2011) 'A parasite in wolf's clothing: hawk mimicry reduces mobbing of cuckoos by hosts'. *Behavioral Ecology* 22: 574–9.

8. Brooke, M. de L. and Davies, N.B. (1988) 'Egg mimicry by cuckoos *Cuculus canorus* in relation to discrimination by hosts'. *Nature* 335: 630–2; Davies, N.B. and Brooke, M. de L. (1989) 'An experimental study of co-evolution between the cuckoo, *Cuculus canorus*, and its hosts. I. Host egg discrimination'. *Journal of Animal Ecology* 58: 207–24; and Davies, N.B. and Brooke, M. de L. (1989) 'An experimental study of co-evolution between the cuckoo, *Cuculus canorus*, and its hosts. II. Host egg markings, chick discrimination and general discussion'. *Journal of Animal Ecology* 58: 225–36.

9. Stoddard, M.C. and Stevens, M. (2010) 'Pattern mimicry of host eggs by the common cuckoo, as seen through a bird's eye'. *Proceedings of the Royal Society of London B: Biological Sciences* 277: 1387–93; and Stoddard, M.C. and Stevens, M. (2011) 'Avian vision and the evolution of egg color mimicry in the common cuckoo'. *Evolution* 65: 2004–13.

10. Spottiswoode, C.N. and Stevens, M. (2010) 'Visual modeling shows that avian host parents use multiple visual cues in rejecting parasitic eggs'. *Proceedings of the National Academy of Sciences of the USA* 107: 8672–6; and Spottiswoode, C.N. and Stevens, M. (2011) 'How to evade a coevolving brood parasite: egg discrimination versus egg variability as host defences'. *Proceedings of the Royal Society of London B: Biological Sciences* 278: 3566–73.

11. For example: Rothstein, S.I. (1974) 'Mechanisms and avian egg recognition: possible learned and innate factors'. *The Auk* 91: 796–807; and Rothstein, S.I. (1975) 'Mechanisms of avian egg-recognition: do birds know their own eggs?' *Animal Behaviour* 23: 268–78.

12. Stevens, M., Troscianko, J., and Spottiswoode, C.N. (2013) 'Repeated targeting of the same hosts by a brood parasite compromises host egg rejection'. *Nature Communications* 4: 2475.

13. Hoover, J.P. and Robinson, S.K. (2007) 'Retaliatory mafia behaviour by a parasitic cowbird favors host acceptance of parasitic eggs'. *Proceedings of the National Academy of Sciences of the USA* 104: 4479–83.

14. Langmore, N.E., Hunt, S., and Kilner, R.M. (2003) 'Escalation of a coevolutionary arms race through host rejection of brood parasitic young'. *Nature* 422: 157–60.

15. Sato, N.J., Tokue, K., Noske, R.A., Mikami, O.K., and Ueda, K. (2010) 'Evicting cuckoo nestlings from the nest: a new anti-parasite behaviour'. *Biology Letters* 6: 67–9.

16. Langmore, N.E., Stevens, M., Maurer, G., Heinsohn, R., Hall, M.L., et al. (2011) 'Visual mimicry of host nestlings by cuckoos'. *Proceedings of the Royal Society B: Biological Sciences* 278: 2455–63.

17. Colombelli-Négrel, D., Hauber, M.E., Robertson, J., Sulloway, F.J., Hoi, H., et al. (2012) 'Embryonic learning of vocal passwords in superb fairy-wrens reveals intruder cuckoo nestlings'. *Current Biology* 20: 2155–60.

18. Spottiswoode, C.N. and Koorevaar, J. (2012) 'A stab in the dark: chick killing by brood parasitic honeyguides'. *Biology Letters* 8: 241–4.

19. Kilner, R.M., Madden, J.R., and Hauber, M.E. (2004) 'Brood parasitic cowbird nestlings use host young to procure resources'. *Science* 305: 877–9.

20. Davies, N.B., Kilner, R.M., and Noble, D.G. (1998) 'Nestling cuckoos, *Cuculus canorus*, exploit hosts with begging calls that mimic a brood'. *Proceedings of the Royal Society of London B: Biological Sciences* 265: 673–8.

21. Work shortly after by the same team partially modified the idea of mimicking a whole brood, and instead suggested that the elaborate begging call of the cuckoo chick might also be explained by the cuckoo needing to beg with an intense call to suggest they are in significant need of provisioning (i.e. very hungry): Kilner, R.M., Noble, D.G., and Davies, N.B. (1999) 'Signals of need in parent–offspring communication and their exploitation by the common cuckoo'. *Nature* 397: 667–72.

22. Tanaka, K., Morimoto, G., Stevens, M., and Ueda, K. (2011) 'Rethinking visual super-normal stimuli in cuckoos: visual modeling of host and parasite signals'. *Behavioral Ecology* 22: 1012–19.

23. See: Brandt, M., Foitzik, S., Fischer-Blass, B., and Heinze, J. (2005) 'The coevolutionary dynamics of obligate ant social parasite systems—between prudence and antagonism'. *Biological Reviews* 80: 251–67; and Kilner, R.M. and Langmore, N.E. (2011) 'Cuckoos versus hosts in insects and birds: adaptations, counter-adaptations and outcomes'. *Biological Reviews* 86: 836–52.

24. Kaib, M., Heinze, J., and Ortius, D. (1993) 'Cuticular hydrocarbon profiles in the slave-making ant *Harpagoxenus sublaevis* and its hosts'. *Naturwissenschaften* 80: 281–5.

25. Regnier, F.E. and Wilson, E.O. (1971) 'Chemical communication and "propaganda" in slave-maker ants'. *Science* 172: 267–9.

26. Allies, A.B., Bourke, A.F.G., and Franks, N.R. (1986) 'Propaganda substances in the cuckoo ant *Leptothorax kutteri* and the slave-maker *Harpagoxenus sublaevis*'. *Journal of Chemical Ecology* 12: 1285–93.

27. Brandt, M., Heinze, J., Schmitt, T., and Foitzik, S. (2005) 'A chemical level in the coevolutionary arms race between an ant social parasite and its hosts'. *Journal of Evolutionary Biology* 18: 576–86.

28. Achenbach, A. and Foitzik, S. (2009) 'First evidence for slave rebellion: enslaved ant workers systematically kill the brood of their social parasite *Protomognathus americanus*'. *Evolution* 63: 1068–75; and Achenbach, A., Witte, V., and Foitzik, S. (2010) 'Brood

exchange experiments and chemical analyses shed light on slave rebellion in ants'. *Behavioral Ecology* 21: 948–56.

29. Foitzik, S., DeHeer, C.J., Hunjan, D.N., and Herbers, J.M.(2001) 'Coevolution in host–parasite systems: behavioural strategies of slave-making ants and their hosts'. *Proceedings of the Royal Society of London Series B: Biological Sciences* 268: 1139–46.

30. Bogusch, P., Kratochvil, L., and Straka, J. (2006) 'Generalist cuckoo bees (Hymenoptera: Apoidea: *Sphecodes*) are species-specialist at the individual level'. *Behavioral Ecology and Sociobiology* 60: 422–9.

31. Martin, S.J., Carruthers, J., Williams, P., and Drijfhout, F.P. (2010) 'Host specific social parasites (*Psithyrus*) indicate chemical recognition system in bumblebees'. *Journal of Chemical Ecology* 36: 855–63.

32. Matsuura, K. (2006) 'Termite-egg mimicry by a sclerotium-forming fungus'. *Proceedings of the Royal Society B: Biological Sciences* 22: 1203–9.

33. Sato, T. (1986) 'A brood parasitic catfish of mouthbrooding cichlid fishes in Lake Tanganyika'. *Nature* 323: 58–9.

34. Stauffer, J.R. and Loftus, W.F. (2010) 'Brood parasitism of a bagrid catfish (*Bagrus meridionalis*) by a clariid catfish (*Bathyclarias nyasensis*) in Lake Malaŵi, Africa'. *Copeia* 2010: 71–4.

CHAPTER 8

1. Brodmann, J., Twele, R., Francke, W., Yi-bo, L., Xi-qiang, S., et al. (2009) 'Orchid mimics honey bee alarm pheromone in order to attract hornets for pollination'. *Current Biology* 19: 1368–72.

2. For further information, see: Gaskett, A.C. (2011) 'Orchid pollination by sexual deception: pollinator perspectives'. *Biological Reviews* 86: 33–75.

3. Schiestl, F.P., Peakall, R., Mant, J.G., Ibarra, F., Schulz, C. et al. (2003) 'The chemistry of sexual deception in an orchid–wasp pollination system'. *Science* 302: 437–8.

4. Stökl, J., Brodmann, J., Dafni, A., Ayasse, M., and Hansson, B.S. (2011) 'Smells like aphids: orchid flowers mimic aphid alarm pheromones to attract hoverflies for pollination'. *Proceedings of the Royal Society of London B: Biological Sciences* 278: 1216–22.

5. Vereecken, N.J. and Schiestl, F.P. (2008) 'The evolution of imperfect floral mimicry'. *Proceedings of the National Academy of Sciences of the USA* 105: 7484–8.

6. Gaskett, A.C. and Herberstein, M.E. (2010) 'Colour mimicry and sexual deception by tongue orchids (*Cryptostylis*)'. *Naturwissenschaften* 97: 97–102.

7. Newman, E., Anderson, B., and Johnson, S.D. (2012) 'Flower colour adaptation in a mimetic orchid'. *Proceedings of the Royal Society B: Biological Sciences* 279: 2309–13.

8. Ren, Z.-X., Li, D.-Z., Bernhardt, P., and Wang, H. (2011) 'Flowers of *Cypripedium fargesii* (Orchidaceae) fool flat-footed flies (Platypezidae) by faking fungus-infected foliage'. *Proceedings of the National Academy of Sciences of the USA* 108: 7478–80.

9. Ellis, A.G. and Johnson, S.D. (2010) 'Floral mimicry enhances pollen export: the evolution of pollination by sexual deceit outside of the Orchidaceae'. *American Naturalist* 176: E143-E151.

10. De Jager, M.L. and Ellis, A.G. (2014) 'Costs of deception and learned resistance in deceptive interactions'. *Proceedings of the Royal Society B: Biological Sciences* 281: 2861.

11. Oliveira, A.G., Stevani, C.V., Waldenmaier, H.E., Viviani, V., Emerson, J.M., et al. (2015) 'Circadian control sheds light on fungal bioluminescence'. *Current Biology* 25: 964–8.

12. Darwin, C.R. (1871) *The Descent of Man and Selection in Relation to Sex*. London: John Murray.

13. Endler, J.A., Endler, L.C., and Doerr, N.R. (2010) 'Great bowerbirds create theaters with forced perspective when seen by their audience'. *Current Biology* 20: 1679–84.

14. Kelley, L.A. and Endler, J.A. (2012) 'Male great bowerbirds create forced perspective illusions with consistently different individual quality'. *Proceedings of the National Academy of Sciences of the USA* 109: 20980–5.

15. Kelley, L.A. and Endler, J.A. (2012) 'Illusions promote mating success in great bowerbirds'. *Science* 335: 335–8.

16. Doerr, N.R. and Endler, J.A. (2015) 'Illusions vary because of the types of decorations at bowers, not male skill at arranging them, in great bowerbirds'. *Animal Behaviour* 99: 73–82.

17. Gasparini, C., Serena, G., and Pilastro, A. (2013) 'Do unattractive friends make you look better? Context-dependent male mating preferences in the guppy'. *Proceedings of the Royal Society B: Biological Sciences* 280: 3072.

18. Gori, S., Agrillo, C., Dadda, M., and Bisazza, A. (2014) 'Do fish perceive illusory motion?' *Scientific Reports* 4: 6443.

19. Hughes, K.D., Higham, J.P., Allen, W.L., Elliot, A.J, and Hayden, B.Y. (2015) 'Extraneous red drives female macaques' gaze toward photographs of male conspecifics'. *Evolution and Human Behavior* 36: 25–31.

20. Fernandez, A.A. and Morris, M.R. (2007) 'Sexual selection and trichromatic color vision in primates: statistical support for the preexisting-bias hypothesis'. *American Naturalist* 170: 10–20.

21. Ryan, M.J., Fox, J.H., Wilczynski, W., and Rand, A.S. (1990) 'Sexual selection for sensory exploitation in the frog *Physalaemus pustulosus*'. *Nature* 343: 66–7.

22. Basolo, A.L. (1995) 'Phylogenetic evidence for the role of a pre-existing bias in sexual selection'. *Proceedings of the Royal Society of London B: Biological Sciences* 259: 307–11.

23. Christy, J.H. (1995) 'Mimicry, mate choice, and the sensory trap hypothesis'. *American Naturalist* 146: 171–81.

24. Christy, J.H., Backwell, P.R.Y., and Schober, U. (2003) 'Interspecific attractiveness of structures built by courting male fiddler crabs: experimental evidence of a sensory trap'. *Behavioral Ecology and Sociobiology* 53: 84–91.

25. Proctor, H.C. (1991) 'Courtship in the water mite *Neumania papillator*: males capitalize on female adoptions for predation'. *Animal Behaviour* 42: 589–98; and Proctor, H.C. (1992) 'Sensory exploitation and the evolution of male mating behaviour: a cladistic test using water mites (Acari: Parasitengona)'. *Animal Behaviour* 44: 745–52.

26. Garcia, C.M. and Lemus, Y.S. (2012) 'Foraging costs drive female resistance to a sensory trap'. *Proceedings of the Royal Society B: Biological Sciences* 279: 2262–8.

27. Nakano, R., Takanashi, T., Skals, N., Surlykke, A., and Ishikawa, Y. (2010) 'To females of a noctuid moth, male courtship songs are nothing more than bat echolocation calls'. *Biology Letters* 6: 582–4.

28. Nakano, R., Ihara, F., Toyama, M., and Toda, S. (2014) 'Double meaning of courtship song in a moth'. *Proceedings of the Royal Society B: Biological Sciences* 281: 0840.

29. Dominey, W.J. (1980) 'Female mimicry in male bluegill sunfish—a genetic polymorphism?' *Nature* 284: 546–8.

30. Gross, M.R. and Charnov, E.L. (1980) 'Alternative male life histories in bluegill sunfish'. *Proceedings of the National Academy of Sciences of the USA* 77: 6937–40.

31. See: Mank, J.E. and Avise, J.C. (2006) 'Comparative phylogenetic analysis of male alternative reproductive tactics in ray-finned fishes'. *Evolution* 60: 1311–16.

32. Jukema, J. and Piersma, T. (2006) 'Permanent female mimics in a lekking shorebird'. *Biology Letters* 2: 161–4.

33. Hall, K.C. and Hanlon, R.T. (2002) 'Principal features of the mating system of a large spawning aggregation of the giant Australian cuttlefish *Sepia apama* (Mollusca: Cephalopoda)'. *Marine Biology* 140: 533–45.

34. Andres, J.A., Sanchez-Guillen, R.A., and Rivera, C. (2002) 'Evolution of female colour polymorphism in damselflies: testing the hypotheses'. *Animal Behaviour* 63: 677–85.

35. Huang, S.-C. and Reinhard, J. (2012) 'Color change from male-mimic to gynomorphic: a new aspect of signaling sexual status in damselflies (Odonata: Zygoptera)'. *Behavioral Ecology* 23: 1269–75.

36. DeMartini, D.G., Ghoshal, A., Pandolfi, E., Weaver, A.T., Baum, M., et al. (2013) 'Dynamic biophotonics: female squid exhibit dimorphic tunable leucophores and iridocytes'. *Journal of Experimental Biology* 216: 3733–41.

CHAPTER 9

1. Wang, M., Bethoux, O., Bradler, F.M., Jacques, F.M.B., Cui, Y., et al. (2014) 'Under cover at pre-angiosperm times: a cloaked phasmatodean insect from the early Cretaceous Jehol biota'. *PLoS ONE* 9: e91290.

2. This idea is somewhat speculative and the authors also discuss a range of other potential explanations for the morphology of the animals, including sensory apparatus. Topper, T.P., Strotz, L.C., Holmer, L.E., Zhang, Z., Tait, N.N., et al. (2015) 'Competition and mimicry: the curious case of chaetae in brachiopods from the middle Cambrian Burgess Shale'. *BMC Evolutionary Biology* 15: 42.

3. Wunderlich, J. (2000) 'Ant mimicry by spiders and spider-mite interactions preserved in Baltic amber (Arachnida: Acari, Araneae)'. *European Arachnology* 2000: 355–8.

4. Miya, M., Pietsch, T.W. and Orr, J.W. (2010) 'Evolutionary history of anglerfishes (Teleostei: Lophiiformes): a mitogenomic perspective'. *BMC Evolutionary Biology* 10: 58.

5. Goldschmidt, R. (1940) *The Material Basis of Evolution*. New Haven, CT: Yale University Press. Goldschmidt agreed that small incremental changes were important in driving evolution, but argued that these only explained small differences between individuals and species. He suggested that large changes through macro-mutations could allow substantial changes between species. While he was ridiculed at the time by many, modern genetics and developmental work has shown that in some cases changes to one or a few select genes can sometimes produce substantial changes in individual appearance.

6. Wallace, A.R. (1867) 'Mimicry and other protective resemblances among animals'. *Westminster Review* (London ed.) 1 July: 1–43.

7. Suzuki, T.K. (2013) 'Modularity of a leaf moth-wing pattern and a versatile characteristic of the wing-pattern ground plan'. *BMC Evolutionary Biology* 13: 158; and Suzuki, T.K., Tomita, S., and Sezutsu, H. (2014) 'Gradual and contingent evolutionary emergence of leaf mimicry in butterfly wing patterns'. *BMC Evolutionary Biology* 14: 229.

8. Spottiswoode, C.N. and Stevens, M. (2012) 'Host–parasite arms races and rapid changes in bird egg appearance'. *American Naturalist* 179: 633–48.

9. Kraemer, A.C., Serb, J.M., and Adams, D.C. (2015) 'Batesian mimics influence the evolution of conspicuousness in an aposematic salamander'. *Journal of Evolutionary Biology* 28: 1016–23.

10. Herberstein, M.E., Baldwin, H.J., and Gaskett, A.C. (2014) 'Deception down under: is Australia a hot spot for deception?' *Behavioral Ecology* 25: 1–5.

FURTHER READING

This is by no means a definitive list, but a selection of books (mostly for a general audience) that readers may wish to consult for other discussions of some of the main ideas I have written about.

Davies, N.B. (2015) *Cuckoo: Cheating by Nature*. London: Bloomsbury Publishing.
  A book about the biology of the common cuckoo and the work done by the author and others to understand it.

Diamond, J. and Bond, A.B. (2014) *Concealing Coloration in Animals*. Cambridge, MA: Belknap Press.
  A short book about camouflage in nature.

Forbes, P. (2009) *Dazzled and Deceived: Mimicry and Camouflage*. New Haven, CT: Yale University Press.
  A book about camouflage and mimicry, focussing especially on military aspects of dazzle camouflage and on some forms of mimicry in nature where two toxic species mimic each other (a type of mimicry not discussed in this book because it does not involve deception).

Smith, C.H. and Beccaloni, G., editors (2010) *Natural Selection and Beyond: The Intellectual Legacy of Alfred Russel Wallace*. Oxford: Oxford University Press.
  A collection of essays about Wallace, ranging from his theories of animal coloration and evolution, through to his ideas regarding society and spiritualism.

Slotten, R.A. (2006) *The Heretic in Darwin's Court: The Life of Alfred Russel Wallace*. New York: Columbia University Press.
  An engaging and detailed biography of Wallace's life and travels.

INDEX

Numbers in italics refer to figures.

THE ANIMAL KINGDOM: A VERY SHORT INTRODUCTION

Peter Holland

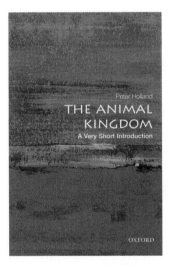

978-0-19-959321-7 | Paperback | £7.99

The animal world is immensely diverse, and our understanding of it has been greatly enhanced by analysis of DNA and the study of evolution and development ('evo-devo'). In this *Very Short Introduction* Peter Holland presents a modern tour of the animal kingdom.

Beginning with the definition of animals (not obvious in biological terms), he takes the reader through the high-level groupings of animals (phyla) and new views on their evolutionary relationships based on molecular data, together with an overview of the biology of each group of animals.

THE DARWINIAN TOURIST

Viewing the world through evolutionary eyes

Christopher Wills

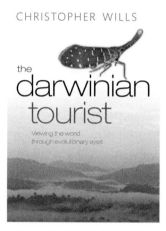

978-0-19-958438-3 | Hardback | £18.99

'Wills' has written a glorious examination of Earth's lesser known biological fauna and flora.' Dave Partridge, *Real Travel*

'There is much to fascinate.' *Times Higher Education Supplement*

In this magnificently illustrated book, Christopher Wills takes us on a series of adventures. From the underwater life of Indonesia's Lambeh Strait to a little valley in northern Israel, each chapter features a different location and brings out a different and important message. With the author's own stunning photographs of the wildlife he discovered on his travels, he draws out the evolutionary stories behind the wildlife and shows how our understanding of the living world can be deepened by a Darwinian perspective. Wills demonstrates how looking at the world with evolutionary eyes leaves us with a renewed sense of wonder about life's astounding present-day diversity, along with an appreciation of our evolutionary history.

Sign up to our quarterly e-newsletter **http://academic-preferences.oup.com/**

A DICTIONARY OF ANIMAL BEHAVIOUR

David McFarland

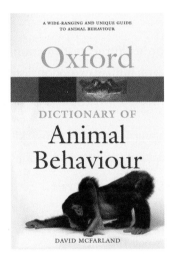

A WIDE-RANGING AND UNIQUE GUIDE
TO ANIMAL BEHAVIOUR

Oxford

DICTIONARY OF
Animal
Behaviour

DAVID MCFARLAND

978-0-19-860721-2 | Paperback | £12.99

Did you know that chickens have different alarm calls for different predators?

This fascinating dictionary covers every aspect of animal behaviour and includes terms from the related fields of ecology, physiology, and psychology. Clear, concise entries are backed up by specific examples where appropriate, covering all aspects of behaviour from aggression to courtship.

The author, David McFarland, was formerly head of the Animal Behaviour Research Group at the University of Oxford. Jargon free and informative, this dictionary is an excellent source of reference for students of biology and psychology, and essential reading for naturalists, bird-watchers, and the general reader with an interest in animal behaviour.

Sign up to our quarterly e-newsletter **http://academic-preferences.oup.com/**

ARE DOLPHINS REALLY SMART?

The mammal behind the myth

Justin Gregg

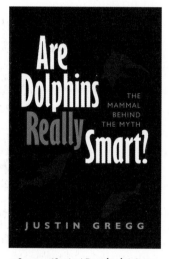

978-0-19-968156-3 | Paperback | £9.99

'Serves as both a rigorous litmus test of animal intelligence and a check on human exceptionalism.'
Bob Grant, *The Scientist*

'[T]horough and engaging [Gregg's] writing skills are solid and his observations are often fascinating.'
Booklist

The Western world has had an enduring love affair with dolphins since the early 1960s, with fanciful claims of their 'healing powers' and 'super intelligence'. Myths and pseudoscience abound on the subject. Justin Gregg weighs up the claims made about dolphin intelligence and separates scientific fact from fiction. He puts our knowledge about dolphin behaviour and intelligence into perspective, with comparisons to scientific studies of other animals, especially the crow family and great apes.